GERM-LINE INTERVENTION AND OUR RESPONSIBILITIES TO FUTURE GENERATIONS

Philosophy and Medicine

VOLUME 55

The titles published in this series are listed at the end of this volume.

GERM-LINE INTERVENTION AND OUR RESPONSIBILITIES TO FUTURE GENERATIONS

Edited by

EMMANUEL AGIUS

Moral Philosophy & Theology, University of Malta
Coordinator, Future Generations Programme, University of Malta

SALVINO BUSUTTIL

Foundation for International Studies, University of Malta

in collaboration with

TAE-CHANG KIM

Institute for Integrated Study of Future Generations, Kyoto, Japan

KATSUHIKO YAZAKI

Future Generations Alliance Foundation, Kyoto, Japan

KLUWER ACADEMIC PUBLISHERS

DORDRECHT / BOSTON / LONDON

A C.I.P Catalogue record for this book is available from the Library of Congress.

ISBN 0-7923-4828-1

Published by Kluwer Academic Publishers
PO Box 17, 3300 AA Dordrecht, The Netherlands

Sold and distributed in the United States and Canada
by Kluwer Academic Publishers,
PO Box 358, Accord Station, Hingham, MA 02018-0358, USA

In all other countries, sold and distributed
by Kluwer Academic Publishers,
PO Box 322, 3300 AH Dordrecht, The Netherlands

Printed on acid-free paper

Printed in Great Britain.

TABLE OF CONTENTS

FOREWORD

As early as 1930, Aldous Huxley (*Brave New World*) and C.S. Lewis (*The Abolition of Man*) predicted that the greatest threat to humanity was not nuclear destruction, but developments in genetics, molecular biology, and biotechnology. Their statements were indeed prophetic!

The rediscovery of Gregor Johann Mendel's laws of heredity in the early part of this century helped promote an understanding of the origins and workings of genetic diversity. The principal phenomena involved are segregation, mutation, and recombination of genes. Together, these three actions, through the opportunities they generate for genetic diversity, have since been used to improve plants, animals, and micro-organisms of interest to agriculture, industry, and medicine. From the beginning of this century, techniques such as planned hybridization and later, the induction of mutations, have been applied by agricultural and biological research institutions to create new genetic combinations. New antibiotics and vaccines came into use and fermentation techniques made rapid strides. Hybridization also became a method of increasing the growth of crops and animals, a phenomenon known as hybrid vigour.

Progress in the field has since been marked by increasingly striking discoveries. The unravelling of the double helix structure of the deoxyribonucleic acid molecule (DNA) inaugurated the era of molecular biology and opened up a new world of genetic engineering. Recent developments have made it possible to identify specific sequences of DNA that are associated with individual traits of organisms. Begun in 1990, the Human Genome Project is an effort to decipher the complete genetic code of the human species. It involves both mapping (locating) all the genes in human DNA and sequencing (determining the order of nucleotides) for each of these genes. Its results are emerging piece by piece as the coordinated efforts of researchers proceed, and many of them – such as the identification of genes for colon cancer and cystic fibrosis – are already beginning to be applied in practice.

While the potential benefits of the Human Genome Project to society are enormous, there are also serious risks from the unanticipated consequences of this powerful new knowledge. The development of routine tests for predispositions to diseases and other human traits raise a host of

E. Agius and S. Busuttil (eds.), Germ-Line Intervention and our Responsibilities to Future Generations, vii–viii.
© 1998 *Kluwer Academic Publishers. Printed in Great Britain.*

complex ethical and legal questions. In short, genetic enhancement risks violating human dignity by opening up the possibility of discriminatory practices. To make sure that this new power does not rebound against humanity, and that scientific and technological progress, source of well-being for individuals and nations, would not be used for self-destructive means, constant efforts must be made not to transgress human rights and freedoms.

The concept of human dignity is decisive here as fresh dangers loom on the horizon. With regard to 'dignity' one recalls the Enlightenment principle that a human being is to be treated as an end, not a means to a further end. The modern humanist understanding of dignity is in part a secularization of the previous religious commitment to the infinite value of the human being. What this means is that a human person is the locus and end of moral value, not to be subordinated to other values presumed to be higher. For this reason, it is not only legitimate but imperative for the international community to do all it can to establish a collective system to defend this essential basic value – the integrity of the human being. The formal inclusion of the human genome in the common heritage of humanity, in other words the recognition that it constitutes one of humanity's primary "resources" which it must safeguard, would undeniably contribute to this end.

The human genome, like any other component of the common heritage of humanity (the seabed, the oceans, celestial bodies), must be subject to the rule of the principle of equality and non-discrimination as regards its use. From this it follows that any advance in knowledge about the human genome must benefit humankind as such. Likewise, in my view, there can be no private rights regarding the human genome, which is not open to appropriation by anyone.

In 1953 Bertrand Russell wrote: "The human race has survived hitherto owing to ignorance and incompetence; but, given knowledge and competence combined with folly, there can be no certainty of survival. Knowledge is power, but it is power for evil just as much as for good. It follows that, unless men increase in wisdom as much as in knowledge, increase of knowledge will be an increase of sorrow."[1] To the perilous leaps in power, associated with war and ecology, we must now add genetic knowledge. Past genetic theories, usually infected with prejudice, have brought the world much sorrow. A new ethical imagination that is slowly emerging through frank debate on human dignity allows us to hope.

REFERENCE

1. Russell, B.: 1953, *The Impact of Science on Society*, Simon and Schuster, New York, 97–98.

EMMANUEL AGIUS

INTRODUCTION

In contemporary ethical discussion about the impact of modern technology and its potential long-range consequences, one particular criterion is gaining more and more prominence. This criterion concerns the extent to which present human activity is seen to affect the well-being and quality of life of future generations. During the last few decades, the ecological crisis has already pointed out the urgent need of technological assessment. For it has become increasingly apparent that although pollution and consumption of resources have local sources, they tend to have not only global consequences but also create future risks and burdens. Moreover, the immense new and large-scale threats posed by both peaceful and military application of nuclear energy have also awakened a greater sense of responsibility for defenceless posterity.

Certainly, today's widespread concern about the potential risks of genetic engineering is raising new and fundamental questions about our ethical relations to unborn generations. The newly acquired knowledge about the mysteries of life and the ability to manipulate the evolution of the human species have immensely increased our power over the future of humankind. It has now become possible to alter life so that it not only affects the present, but also the gene pool of all future generations as well. C.S. Lewis says that "each new power won by man is a power over man as well." Indeed, the ability to penetrate the nucleus of the living cell, to rearrange and transplant the molecules of nucleic acid that constitutes the genetic material of all forms of life has brought about not only potential benefits but also immense risks to both present and future generations.

The recent rapid expansion of research in the area of molecular genetics, resulting in large part from the application of recombinant-DNA technology, has vastly increased the possibilities to determine the genetic make-up of an individual both before and after birth. Genetic engineering techniques are already demonstrating their great potential value for human well-being. During the last decades, the prospect of direct application of gene splicing to cure human genetic diseases has moved forward by large steps, though great difficulties still have to be overcome.

Although gene splicing techniques, particularly germ-line and zygote

E. Agius and S. Busuttil (eds.), Germ-Line Intervention and our Responsibilities to Future Generations, ix–xiv.
© 1998 *Kluwer Academic Publishers. Printed in Great Britain.*

therapies, hold great promise for the future, technical problems and uncertainties prohibit for the time being their deliberate application. It is irresponsible to proceed while results are uncertain since technical failures might endanger unborn generations. To use such therapy, while not yet perfected, would risk creating offspring who have genetic problems created by the therapy rather than naturally occurring defects. Despite the enormous amount of information obtained on DNA in recent years, our basic understanding of cells and how they function is still exceedingly elementary. Moreover, another technical drawback at present is that transferred genes integrate randomly in the genome. Methods for the reliable introduction of functioning genes into germ cells and for verifying their successful introduction still need to be developed.

In addition to the technical uncertainties involved, the misuse of DNA technology for non-therapeutic purposes raises serious ethical questions concerning our relationship to posterity. Gene splicing techniques can be used for "positive" eugenics in order to change the basic characteristics of human nature rather than to cure chromosomal disorders. Moreover, it can become a tool of tyrannical malevolence used to manipulate human beings for political and social purposes.

This book raises a number of ethical issues about the impact of genetic engineering on generations yet to be born. The four topical areas that constitute the focus of this volume were chosen because they raise intriguing ethical and legal questions, as well as important policy issues. As much as any set of issues, they reflect the hopes, fears, prejudices, and uncertainties that people associate with germ-line intervention and the future of humankind. The essays attempt to provide empirical information accompanied by a moral framework for deliberations on the issues involved in genetic engineering and the Human Genome Project at large.

While the potential benefits of the Human Genome Project to society are enormous, there are also serious risks from the unanticipated consequences of this powerful new knowledge. The development of routine tests for predispositions to diseases and other human traits has profound implications for medical practice, for insurance, for the legal system, and the well-being of our progeny. Section One, 'From Laboratory to Germ-line Therapy', aims at helping society reap the benefits of the new genetic knowledge while addressing in a constructive manner the potential dilemmas created.

Alfred Cuschieri expounds the thesis that the greatest contribution we can make to future generations lies in the present, for it is the present genes that can be modified, in a beneficial or detrimental way, by today's geneticists. He also stresses the moral obligation that genetic testing brings with it, namely the responsibility to provide adequate supporting and counselling services for those availing themselves of genetic tests. He also underscores the need to educate public opinion in order to dispel undue fears.

Wilai Noonpakdee explores 'The Moral Status of the Human Genome' and highlights the different approaches that ethicists take in pondering the relationship between deterministic theories of human behavior and philosophical concerns for agency, individuality, and morality. She also raises the problem of whether the focus in gen-ethical thinking should be on the relative privacy of the individual or the collective whole.

The concept of human nature and different ways of grounding moral constraints in genetic therapy are discussed in Section Two. Kido Inoue, a Buddhist priest, explores the relationship between our spiritual genes and the surrounding environment which in his view influences greatly character formation and the propagation of sound genes to our progeny. He sees the disciplining of our primitive instincts and the transcending of our egoistic drives as necessary in creating and sustaining a healthy natural environment that helps individuals go beyond their egos and reach out to others in love and compassion.

Salvatore Privitera examines the application of teleological and deontological moral reasoning to genetic engineering. While he sees both approaches as contributing to a better understanding of the bioethical debate, in his view the teleological approach accommodates best the problematic issues that genetic engineering raises with regard to future generations.

Kevin Wildes advances the claim that each of the general philosophical systems fails to generate general moral constraints apart from consent, for the general secular context is committed to no particular set of moral values or justification. He argues that an appeal to human nature, as such, does not support moral constraints for a secular, morally pluralistic society, for there is no one fixed identifiable human nature that could be morally normative. To discover more robust, content-full constraints one must turn to moral communities where one learns what are the moral virtues that are to be admired and vices that are to be disdained. But no one set of values shared by all men and women could ever be defined. Likewise, there will be no general secular view of how genetic therapies ought to be used. Indeed, because there are competing views of human nature there will be competing views of what constitutes health and disease that, in turn, yield different views of appropriate medical therapy.

Section Two closes with a contribution from Tristram Engelhardt. Engelhardt sees no absolute basis for rejecting the alteration of the human genome. For him, human freedom and moral diversity go hand in hand and tolerance must accommodate the numerous yet different approaches. Attempts to provide content-full international accords in this area are doomed to failure for there is no one single vision of the goals that should guide genetic interventions. At most, in approaching these difficult moral questions one may be guided by three general moral principles, namely: avoiding malevolent acts against future generations, refraining from performing changes in the human genome that on good grounds would be unacceptable by the recipients, and acting prudently so as not to cause more harm than benefit. In each case, any content must be provided by those considering the interventions, not those who will be subject to them.

Section Three directs the discussion to 'the common heritage view' that favors the law of solidarity and cooperation over the law of competition and self-interest. Emmanuel Agius addresses the need to alter the current intellectual property system for the regulation of biotechnological interventions, because of the potential threat to the future of the existing species. The widespread practice of patenting genetically engineered life forms threatens the unity of species and can lead to loss of genetic diversity. Hence Agius advocates that the present patent law systems be changed in accordance with the innovative concept of the 'common heritage of mankind' that considers the human genome as property of humankind as a whole, and encourages scientific researchers to operate under the principle of communal ownership of knowledge.

Eric Juengst disputes the assumption that the proper objects of our concern in contemplating human germ-line engineering are the descendants of the engineered. Rather, the most important need will be to protect those among future generations who will be at a disadvantage in living and working with their genetically engineered neighbours. Juengst is of the opinion that obligations to future generations will be met by preventing the practice of germ-line intervention from exacerbating forms of genetic discrimination we already know too well.

The final group of essays provides an overview of the social responsibilities of geneticists towards future generations. Qiu Renzong sets out to explore a new conceptual framework that regards individuals as a community of persons-in-relation rather than separate selves each claiming his or her own human rights. Such an approach would favor the inclusion of future generations who tend to be excluded when one appeals only to the rights of individuals. Addressing the issue of eugenics, Qiu draws a moral borderline between genetic intervention for treatment and interventions for enhancement, and categorically rejects positive eugenics due to unforeseen, negative consequences.

Alex Felice reminds us of the value of time in passing value judgments on the advances in genetic engineering. A review of past and present experiences provides guidelines in making projections for the future. Widened peer review groups provide assurance that good science is adequately and stably supported while guarding the powerless, future generations. This is why, according to Felice, a guardian might function on the global level in integrating and sustaining the efforts of local and institutional peer review groups.

The metaphor of a guardian within the context of genetic engineering is explored by David Heyd. He argues that founding responsibility in genetic policies on the idea of guardianship is fraught with problems, for rights are derived from interests, and interests are a function of what the subject of the interest is. Hence the fact that future generations lack an independent identity equally undermines both the attempt to extend the principles of justice to genetic choices as well as the idea of nominating a guardian to oversee the genetic welfare of future generations. As an alternative, Heyd proposes that we speak of our obligations rather than their rights.

The final essay in the volume directs our attention to political philosophy. Stuart Spicker attends to the political mechanisms within democratic republics that have thus far been proven efficacious in permitting, encouraging, and funding genetic engineers not only in acquiring knowledge through their research but also in monitoring their accountability to the public. He asserts that prudence calls for minimum rational constraints on geneticists, via democratic political processes, to ensure that they stay true to what Hans Jonas calls "the internal norms of scientific inquiry."

I hope that readers will be stimulated by the ethical issues raised by the contributions to discuss critically and creatively the problems that confront us all. As far as I know, this publication is the first bioethical contribution which deals with genetic engineering from the perspective of our responsibilities towards future generations.

It has been a pleasure to complete this volume which contains papers presented at a three-day symposium held in Malta. This conference would not have been possible without the generous financial support of the Future Generations Alliance Foundation of Kyoto, Japan. Special thanks are due to Katsuhiko Yazaki, Chairman of the Future Generations Alliance Foundation, for his personal interest in this conference and publication. I wish to acknowledge my debt and express my gratitude to the editors of the Philosophy and Medicine series, H. Tristram Engelhardt, Jr., and Stuart F. Spicker, whose counsel and support encouraged me to publish this volume. I express my gratitude to the authors for their cogent contributions. Finally, I thank Stephen Caruana and Lionel Chircop for their painstaking and diligent work in correcting the manuscript and completing the final version.

Future Generations Programme
Foundation for International Studies
University of Malta

PART I

FROM LABORATORY TO GERM-LINE THERAPY

ALFRED CUSCHIERI

SCREENING FOR GENETIC DISEASES: WHAT ARE THE MORAL CONSTRAINTS?

Now that the Human Genome Project (HGP) is an ongoing and rapidly progressing reality, and human genetic engineering is expected to become standard procedure, the inevitable question is how will these procedures be applied. The inevitable and much debated answer is eugenics. It is often looked upon as positive eugenics, directed perhaps, towards achieving human beings endowed with optimal characteristics of physical strength and beauty, intellectual genius and longevity. There is of course the immense and probably insoluble problem of determining which human characteristics, among nature's rich and superb diversity, can be improved and what constitutes the hypothetical physical and intellectual excellence that one might envisage and enhance.

Assuming, purely for the sake of argument, that it would be possible to achieve this excellence, for ourselves and for our future offspring, through genetic engineering, there would immediately arise two significant moral problems. The first would arise from the great likelihood that the vast majority of us would rush to be genetically engineered, creating the moral problem of who should have priority. The second is, that those of us who did not have the privilege of being genetically engineered, or did not have the natural endowment of perfection, would somehow be considered as inferior, second-rate human beings raising the other serious moral worry of discrimination.

Being simply a practicing geneticist who sees an abundance of problems of genetically induced misery, I must confess that my competence in speaking about achieving such perfection is limited. My eugenic priorities dictate that I limit myself to a discussion of the prevention and correction of genetic diseases as a contribution to eugenics for the future. As a geneticist who witnesses some of the most distressing diseases which affect mankind, including the birth of congenitally malformed infants and serious, incapacitating and often lethal genetic diseases; who feels the psychological trauma experienced by parents as a result of the knowledge that one or both of them were agents in the genetic transmission of disease, I cannot help but think that negative eugenics (or the prevention of genetic disease) is, and will remain for a considerable time, the priority

E. Agius and S. Busuttil (eds.), Germ-Line Intervention and our Responsibilities to Future Generations, 3–11.

with respect to future generations. The great strides currently being made in the field of genetic screening is directed precisely towards alerting geneticists to the presence of disease-producing genes as a first step in the prevention of genetic disease and thus contributing to eugenics concerning future generations.

It cannot be denied that the development of human genetics and its remarkable achievements have been spurred on by the relentless quest to find the underlying causes of genetic and inherited diseases and to find remedies for these hitherto untreatable conditions. This is the present ambition of most geneticists and scientists, as can be witnessed by their contributions to the immense literature on genetics. It is a positive contribution: to improve the present gene pool which is the one that will be passed on to future generations.

Given the considerable advances taking place in gene technology, the prospects of genetic engineering and particularly of germ-cell engineering, can we feel confident that we will live up to our responsibilities towards future generations? Is there cause for concern when one considers the prospects of positive eugenics, of creating clones and designer babies? In trying to assess the gravity of the situation one should start by assessing what the present generation of geneticists and scientists are doing and what direction is being taken, because this is what is within our capabilities to oversee. It is our responsibility to guard the present gene pool and ensure, in the most cautious and enlightened way possible, that nothing is done which may be detrimental to future generations, and that necessary measures are taken to implement any positive measures for its enhancement. Should we concern ourselves with what might happen in the future and think of ways in which such eventualities might need to be dealt with?

My thesis is that the greatest positive contribution that we can make to future generations lies in the present. Genes are transmitted one generation at a time. The present genes are the ones which can be modified by the present generation of geneticists and scientists in ways that could be beneficial or detrimental to future generations. How scientists might behave or react in the future is not within our control. However, it is within the responsibility and control of the present generation to criticize the values that emerge from the vast and rapidly expanding field of human genetics and genetic technology, and perhaps to set the trend in the development and evolution of thought regarding future projects.

Consequently, I will not here pursue the philosophical questions that arise should we conduct genetic interventions to enhance desirable characteristics, but I shall dwell on the more mundane and practical issues of what we can do, and what we are doing now, and how we could benefit future generations.

The HGP is the major landmark and turning point of the present time. It is meant to ensure that there is a complete map, catalogue, and sequence of the 100,000 (or probably much more) estimated genes in the human genome, how they function, and how they interact with one another. The project is also expected to provide important information about the large amount of non-coding genetic DNA which apparently does not carry genetic messages in the conventional sense, that is genes for the synthesis of proteins. This vast amount of largely unexplored DNA certainly holds many yet untold secrets about gene interaction and control. The HGP is expected to provide a better understanding of the common heritable traits which do not appear to be attributable to single genes, such as stature, intelligence, obesity, and others. Above all, it is expected to form the basis and the main reference point for investigating inherited and acquired disease, human development, and evolution.

When the HGP is completed we would be in a position to say that we understand human life much better than ever before; that we have a grasp on the mechanisms operating in the human genome. Equipped with all this knowledge man would have control over his own genome and would be able to manipulate it. However, we need not wait until the project is completed. The HGP has already started, and although there is still a long way to go to its full realization, it has already yielded results and its effects are already being felt.

The genes for a number of serious diseases have already been mapped, cloned, and sequenced. This has made possible genetic testing for quite a number of genetic diseases. Genetic testing has already been with us for quite some time. The moral and ethical problems encountered in the course of such testing are therefore not new. However, their magnitude has increased and will continue to increase in proportion to the increase in our knowledge of genes and their mutations.

Advances in genetic technology not only increased the number of tests for genetic diseases, but also made the testing simpler and less costly so that large scale genetic screening has become a possibility and can even be extended to populations.

The greatest moral problems that arise from the ability to test for human genetic disease are (as Baroness Warnock points out) problems of knowledge. Genetic testing is simply the availability of knowledge about the genetic constitution of a particular individual. The problems arise from the implications of such knowledge to the individual concerned, to his or her family, to third parties, and to society in general. Who has the right to such knowledge? Who should decide whether genetic information about an individual is to be obtained or not? That is, who is to decide whether a particular individual should be tested or not? Who should decide to whom this knowledge should be made available? Who should decide what actions are to be taken when a gene abnormality is discovered? These problems are the main determining factors that impose moral constraints on the use of genetic testing in the future.

I begin with the implications of genetic testing for the particular individual. Inherited diseases which manifest themselves late in life are the ones in which these implications are most significant, but they can also be used as models for other diseases. Individuals who are afflicted with severe genetic diseases which appear late in life, such as Huntington's disease, experience the normality of life, the hardship of being afflicted with a serious and incapacitating disease, and the sorrow caused by the daunting prospects of transmitting the same disease to their children and to future generations. Huntington's disease appears late in life, usually after the reproductive years, and often after the individual has transmitted the deleterious genes to the next generation. It is possible to know, through genetic testing, whether a particular individual who is at risk is affected or not, and in my view, the individual should be free to make his or her own choice about whether to be tested, whether he or she would prefer the reality of knowledge to the uncertainty of chance. With the former, distressing as it might be, it would be possible for an individual to plan his or her life, including marriage and children. Some individuals, however, prefer not to be tested, to leave everything to chance, and to live with the uncertainty of risk. The decision is expected to vary depending on the individual and his or her family circumstances. One main factor is whether the individual is already married and has children, or whether he or she is considering marriage or having children. The present trend is that young individuals prefer to "face reality" before entering into relationships with other persons, and that marriage partners expect to know the truth about their future spouse.

Because it is essential to respect an individual's freedom of choice, it is generally accepted that the options for genetic testing are not extended to minors, except in situations where outcomes could be of direct and immediate benefit to a child; as, for example, in a boy with muscular dystrophy or a child with familial intestinal polyposis which requires surveillance for possible malignant growth.

The implications of being tested are serious, and the responsibilities are great, so that the individuals concerned need psychological support and adequate knowledge about the disease itself, the mutant genes and the possible implications, in order to be able to take responsible decisions. In other words, the opportunity for testing brings with it the moral obligation of providing adequate support and counselling services. If these cannot be provided, it seems unethical to carry out genetic testing.

What are the options for preventing the propagation of genetic diseases? One time-honoured option is to voluntarily abstain from having children, a decision which must be entirely free and which should not, in my view, be forced upon prospective parents. Although this approach is usually adopted, couples often do feel the strain of not being able to have children, and typically seek alternatives. The second option is to elect selective abortion following prenatal diagnosis. The third option, which is now becoming more easily available, is pre-implantation diagnosis on a four-\or eight-cell embryo. One of these cells is removed by micro-techniques and its DNA amplified and tested. If it is found to carry the abnormal gene, the remaining embryonic cells are discarded; whereas if the tested cell is found to carry a normal gene, the embryo is implanted. These three methods of inhibiting the propagation of an abnormal gene are all methods of artificial selection against the disease gene.

It is not appropriate to discuss here the moral issues which relate to selective abortion and pre-implantation diagnosis, procedures which some consider morally acceptable but others consider morally unacceptable, and therefore not really options for them. However, irrespective of the acceptability of the procedure itself, one questions whether it is morally right to terminate a pregnancy, whether in the pre- or post- implantation stage, because a disease may appear thirty to sixty years later. Besides, the condition may be treatable. Is it morally justifiable to deny life to an embryo or fetus because it is carrying a gene which predisposes to cancer, manic depressive psychosis or schizophrenia? All of these are undeniably serious diseases, but it should not be concluded that they necessarily make life not worth living. To carry the implications further, is it justifiable to deny life to a fetus because it carries a pre-mutation which does not cause any disease in the individual who carries it, but might manifest itself as a serious disease in future generations, such as may occur in

X-linked mental retardation? These are serious issues which confront us and which require responsible decisions, since they affect future generations.

Besides, the very use of genetic testing and selective termination, rather like weeding a garden of unwanted specimens, is repulsive to many when applied to fellow human beings, and constitutes one of the worst forms of discrimination, imposing therefore one of the most serious moral constraints on the use of genetic testing and subsequent termination of pregnancy .

What are the alternatives? Genetic engineering is expected to provide possible solutions. One of these might be germ-line cell genetic engineering intended to replace the mutant gene with a normal one. Once corrected, the engineered germ-line cell would be expected to ensure that the disease is not transmitted to future generations. However, this approach is replete with uncertainties. Firstly, although a known defective gene may be successfully engineered in a germ-line cell, new mutations still continue to arise so that genetic diseases can only be limited to a certain extent. Secondly, genetics is only beginning to alert us to the fact that genes in germ cells may behave differently from those in somatic cells, and furthermore that they may behave differently in female and in male germ cells. These scientific uncertainties impose a grave moral constraint on the use of such procedures, which should not be employed until there is at least sufficient scientific knowledge about the genetics of germ cells to make germ-line genetic engineering a worthy proposition. Besides, genetic experimentation on germ cells which are subsequently fertilized would pose moral and ethical problems which are in many ways similar to experimentation on embryos. Because of these uncertainties, genetic engineering is, at present, not permissible on human germ cells. Nevertheless, these problems are not unsurmountable, and it is expected that in the future-germ line cell genetic engineering will become feasible and could make important contributions to the prevention of genetic disease.

A feasible alternative to germ-cell engineering is expected to be genetic engineering of the early embryo in its two-\or four-cell stage. Here, one is dealing with somatic cells but the results of the genetic engineering would also include the subsequent germ cells. Genetic engineering in embryos is considered to be a sophisticated and extremely delicate form of surgery intended to correct a defect, and at least prima facie would not appear to present serious moral constraints.

It has already been stated that it is essential to maintain strict privacy in genetic testing especially for adult-onset disorders which are to be performed following the obtaining of informed consent. However, it is rightly argued that the offspring, siblings, and other close relatives of an affected person have a right to know the results of genetic testing, because they too might be at risk. Pressures are also applied by insurance agencies and employers who claim that they have a right to know of existing genetic risks before entering into a contract with their customers. If these demands are acceded to they could easily amount to compulsory testing. These matters, presently strongly debated, are fundamentally concerned with the question of privacy, of who should decide on whether genetic testing is to be performed, and most importantly, with the issue of discrimination which might take the form of increased premiums, of denial of medical or life insurance, or work. Such considerations are more important in some countries where health services rely on private insurance, but of less significance in countries like Malta where full health services are available to all. A similar problem of privacy arises when a population-wide screening program is proposed for a common genetic disorder. This too might amount to compulsory testing.

As genes for more and more diseases become known, and genetic screening becomes easier to perform, the pressures for indirect compulsory screening are bound to increase, though perhaps surreptitiously. It is imperative that such pressures be resisted, not only to guarantee confidentiality of genetic data but above all to ensure that genetic screening is not misused as a form of discrimination.

In assessing the most recent contributions of human genetics, we see that the most prominent advances have been in the field of medical genetics. This is not fortuitous but stems from the real and genuine appreciation of the need to understand and eventually find remedies for serious genetic illnesses, which have for so long eluded us. Medical care has always been committed to preserving life and curing disease. This is not by any means an exclusive characteristic of medical people but merely a commitment to help others, just as people have committed themselves through their various professions to voluntarily care for the weak, the diseased, the disabled, the oppressed, the socially disadvantaged, and those who have been abused by others. It is true that some people have, on the contrary, committed themselves to greed. Why should we now doubt the good intentions of geneticists and scientists who have worked and are still working to identify the basis of genetic disease and in the process are unravelling the fundamental structures of life? Should they be mistrusted and treated as though they were irresponsible?

Most of the concern that has been expressed regarding possible misuse of genetics to the detriment of future generations stems from the fact that exploration of the genome is largely exploration of unknown territory, which always brings with it a certain degree of fear and anxiety. However, danger is not really a deterrent but a challenge to proceed with caution. Many fears and concerns had been expressed in the 1970s when the first genes were cloned in bacteria, but scientists cautiously but steadily carried on in their venture while concerned onlookers cried out DANGER! and even warned of impending doom. Scientists and institutions responsible for their funding took the appropriate measures to ensure that scientists were self-critical and self-controlling. There was consensus on this matter at national and international levels. In 1989, at the Council of Europe, ministers of European countries agreed that "throughout the execution of the programme the ethical, social and legal aspects of human genome analysis should be the subject of wide ranging and in-depth discussions, and possible abuses of the results or later developments of the work should be identified. Principles for their utilization and control should be proposed." It is incumbent on scientists, doctors, and others involved in research to ensure that the application of knowledge is for good purposes and that possible abuses are detected and stopped. Who can do this monitoring better than the scientists themselves? Who knows what is going on and who are in the best position to detect abuses?

Now continuation of genetic research is leading to a better understanding of human life, and ourselves. However, people are again expressing fear and mistrust of science and the powerful technology it creates. Again, this stems largely from fear of the unknown and feelings of threat from a powerful technology which is in the hands of scientists. We hear warnings of another impending calamity (due to the misuse of genetic engineering in human germ cells) posing a threat to the human genome of future generations unless action is taken to prevent it. But, we should stop and reflect. Is it realistic to think that scientists and geneticists involved in this tremendous project are so overwhelmed by the momentous discoveries that a "watchdog" is needed to oversee their activities? Is there really any threat to the genome of future generations, or is it time that we take count of the reality of the situation as it is at present?

Organizations dedicated to protecting future generations usually have good intentions: to promote the view that what man has today (whether this is cultural, environmental, or genetic) should be preserved, and possibly even enriched for future generations. At the moment, we are living up to our responsibilities of preserving the human genome and of passing it on to future generations. We must, however, evaluate the present situation of the human genome, give due recognition to beneficial achievements, and identify deleterious ones, doing all in our power to inhibit them. We must educate the public and dispel undue fears. The future heritage of mankind depends on what we do or fail to do today.

There is no doubt that the most important applications of powerful genetic technology have been in the field of genetic diagnosis, and treatment using genetic engineering is well on the way to new applications. These are the tremendous contributions of medical genetics to the control of human disease. There is no doubt that this ideal leads the list of eugenic priorities. Genetic testing is only one small but essential step toward achieving these priorities. Let us look forward with courage and optimism to achieving additional positive gains which will safeguard and enhance the future of humanity!

Department of Anatomy
University of Malta

THE MORAL STATUS OF THE HUMAN GENOME

I. INTRODUCTION

Ethicists have long pondered the relationship between deterministic scientific theories of human and animal nature and behavior, and the philosophical clarification of concepts – such as morality, agency, and individuality. Plato, in his attempts to construct the morally ideal state (*Republic* III,410,IV pp. 456–61) noted that human beings differ greatly as to their innate natural abilities, that results in the fulfilment of an appropriate function in the state. In other words, there is a correlation between natural abilities and particular functions necessary for harmonious political and moral life. Plato's commitment to such a policy can be seen in his view of the state's role in regulating reproduction. Selective breeding was encouraged as a means of improving society. Strict policies on marriage based on desirable "hereditary" tendencies were introduced as a possible mechanism to achieve such a goal.

II. THE GENERAL CONCEPT OF THE HUMAN GENOME AND ITS IMPORTANCE

The advance of scientific research in the fields of genetics, biochemistry, cell biology and physical chemistry has enabled us to examine genetic materials that reside in each of our cells, controls the development of cells, tissues, organs, our physical traits, and all body functions and activities; these are passed on to future generations. Scientific research aims to reveal the secret of life that ultimately could lead to many changes in our lives and the lives of unborn generations, since we may be able fully to control and manipulate the human genome.

The human genome has a total number of about 3 billions DNA base pairs organized into 23 pairs of chromosomes. However, not all of these sequences contain useful coding information. Only the segment of DNA molecule that contains the code for making protein or enzyme is called a gene, and we have a total of 50,000 – 100,000 genes. These genes are presumed to be arranged in a perfect way. If there is any change in the

E. Agius and S. Busuttil (eds.), Germ-Line Intervention and our Responsibilities to Future Generations, 13–18.

DNA sequence of a gene, abnormal proteins are produced, and our body structure and functions will malfunction leading to thousands of inherited diseases (such as thalassemia and cystic fibrosis) and make us more susceptible to yet uncured diseases, such as heart diseases, Alzheimer's disease, cancer, and some psychiatric illness. These disease-causing genes have been identified. Perhaps no longer then 10 years from now all human gene structures and functions will be elucidated according to the objectives of the multimillion dollar human genome project going on in the United States. This could give us power to control our genes, which will revolutionize medical science in the treatment and diagnosis of diseases, together with understanding human development and even the aging process. At the present time, new knowledge has already led to many ethical dilemmas in applying it in the medical area, not to mention non-medical areas where the manipulation of genes may also be possible.

III. THE HUMAN GENOME PROJECT

The Human Genome Project in general is aimed at the complete mapping and sequencing of the human genome. Such information will facilitate greatly the elucidation of genetic diseases, such as cystic fibrosis. Continually improving methods for mapping genes on chromosomes have offered all of medicine a new paradigm, particularly in the study of the most puzzling diseases, by first mapping genes responsible for them.

The discussion has so far focused on the basic function of science to bring an understanding of nature that it is already there. Usually it will not directly deal with values or ethics. Science has its objective to explain "what is." However, the application of science and technology deals with "what is ethically appropriate?"

Should the state embark on the human genome project? Ramsey, in *Fabricated Man: The Ethics of Genetic Control*, maintained that there is a fundamental obligation and it is morally just to improve the human species. Glass, in *Science and Ethical Values*, also noted that there is a fundamental obligation, in the light of genetic knowledge, not knowingly to produce beings with the prospect of physical or mental defect. Sperry, in *Science and the Problem of Value,* similarly shared the view that man's part is to direct the path of evolution. In other words, to be human means to change nature and the human estate. Moreover, some ethicists hold the position that an empirical understanding of human nature and its biological make up is a necessary prerequisite for constructing political and ethical systems. It was thought that the only plausible social and ethical systems were those that were consistent with scientific facts about human biological nature. Sharing the above mentioned views, the state, then, is obligated to support the human genome program, which can achieve the greatest genetic good. If so, decisions must be taken with regards to funding such a project.

Second, the issue arises of ownership and control of the use of information concerning genetic registries. How can confidentiality be assured especially in an era of computerization? McKusick (1992, pp. 36–42) voiced his concern surrounding the use, misuse, and possible abuse of the information obtained by the Human Genome Project, and the power it generates for discerning the genetic makeup of individuals. Capron (1990, pp. 678–82; 684–96) raised a more subtle issue: the risk that the decision of any person to participate in a family linkage registry may not be truly voluntary, especially since the success of the program rests on the registry of any member of a particular family. The failure to allow oneself to be studied might be regarded as a form of ingratitude, or even as an impediment to the development of a medical response to a disease that burdens the entire family, or even some of its members.

Third, the most potential danger of the human genome project is the ever-increasing tendency toward biological determinism. This worry surrounds the aggressive promotion of genetic testing that may lead to fears out of proportion to the actual risks. Proctor (1992, pp. 89–93) warns of the rush to identify genetic components in cancer or heart disease and worries that the substantial environmental origins of these afflictions may be drastically lessened.

IV. GENETIC TESTING & DIAGNOSIS AND SCREENING

The rapid advances now occurring in genetic testing, diagnosis, and screening, and the increased resources devoted to genetic counselling give the Thais new opportunities to understand their biological heritage and to make their health care and reproductive plans accordingly. One important goal (in Thailand) of genetic diagnosis and counselling is to prevent genetic disease in individuals and families. Generally speaking, the common approach to this problem has been to identify moderate to high risk families (through the birth of an affected child) and then provide detailed genetic counselling to such families. Counsellors usually provide necessary and relevant information of the increased recurrence risk.

This approach to prevention will lessen the number of affected children born. Ideally, it follows, that if any significant reduction in the case load of a particular disorder is expected, couples or individuals at risk will have to be identified before the first affected child is born. In the north-eastern part of Thailand, in particular, it happens to be the case that sickle cell anaemia affects people in the area. So far, in Thailand, reliable treatment for sickle cell disease, or thalassemia, is unavailable and prenatal diagnosis, although possible is quite risky. The power of prevention of the Thalassemia program has prevented a number of such cases. However, a few conflicting ethical issues emerge from this program:

1. the physician's obligation to tell the truth;
2. the patient's right to disclosure;
3. the patient's desire for well being;
4. the patient's occasional desire to remain ignorant of events that might lead to reproductive isolation and/or self-destruction; and
5. finally, concern for the potential health of future offspring and the quality of the "gene pool." By introducing various programs, such as sterilization or rigid restriction of reproduction of those persons at risk to develop Thalassemia, it is likely to reduce it or nearly eliminate it. However, it is interesting to note that the value of eliminating this source of personal suffering must be weighed against and the pain and loss of the thousands of people enrolled in such a program.

It is extremely difficult to find a satisfactory solution. However, difficult ethical issues must be confronted. For example, genetic information provided to relatives or other relevant physicians, in principle, cannot be released without the patient's consent. However, in practice, it is released on the following conditions:

1. reasonable efforts to elicit voluntary consent to disclosure have failed;
2. there is a high probability that harm will occur if the information is withheld, and that identifiable individuals may suffer if the information is withheld.

V. HUMAN GENE THERAPY

L. Walters has raised ethical issues in human gene therapy by classifying four major types of potential genetic intervention with human beings:

Type-1 genetic intervention treats only non-reproductive cells to cure diseases in individual patients. Type-2 genetic intervention affects reproductive cells, and if successful, would reduce the incidence of a particular genetic disease or prevent that genetic disease in the descendants of the treated individuals. Type-3 genetic intervention would enhance selected characteristics in treated individuals only. And Type-4 genetic intervention would pass on such "enhancements" to future generations of a particular family through the germ-line.

In Thailand, human gene therapy is still being considered with regard to the social and ethical implications of this rapidly evolving field. As with any research that involves human subjects, careful attention must be paid to both the immediate and the long-term impacts. Human genome therapy ought to require a continuing public examination of all emerging questions posed by developments and prospects in the human applications of molecular genetics. Our initial response is to assemble a diverse group of consultants that include representatives from medicine and biology, religion, philosophy and ethics, law, and social policy. In other words, the group should be broadly based and not dominated by geneticists, although it should be able to turn to other experts for advice.

As previously mentioned, the Thais are still pondering a number of difficult questions. In particular, seven significant queries appear in "Point to Consider in the Design and Submission of Human Somatic-Cell Gene Therapy Protocols?" (*Federal Register*, no. 160, 1985, pp. 33463–67), are as follows:

1. What is the disease to be treated?
2. What alternative treatments for the disease exist?
3. What are the potential harms of gene therapy for the patients treated?
4. What are the potential benefits for human patients?
5. How will patients be selected in a fair and equitable manner?
6. How will patients' voluntary and informed consent be solicited?
7. How will the privacy of patients and the confidentiality of their medical information be protected?

It is generally accepted that the power of medicine to cure and to prevent illnesses has increased enormously during the past two decades. Most of the advances in medicine have depended on carefully conducted research, supported in large part by public funds. The research, in turn, often depends on the willingness of human volunteers to agree to participate in studies that may present some degree of risk.

In research, benefits to the community and future generation are achieved at the expense of a few. Will Thai society have the obligation to care for or make whole "those who have been injured for the collective good." If not, who should be responsible for the welfare of those who are injured as a result of participating in genetic research? Besides the above noted ethical dilemmas, the Thais are confronting the reality of limited resources. As the capabilities of imported western and medical technologies have expanded, the strain placed on already limited budgets and the government has become a matter of increasing concern.

The reality of scarce resources means patients, health care professionals, institutions, and society as a whole must face an ethical problem – having to make choices within health care budget between treatment and research, between gene therapy as preventive for those who may be at risk and restorative for those already ill. Even though the prospect of human genome therapy for the treatment of disease is comparatively bright, Thai society must ensure that gene therapy is aimed principally at the treatment of disease.

Department of Biochemistry
Mahidol University
Bangkok, Thailand

BIBLIOGRAPHY

Capron, A.M.: 1990, 'Which Ills to Bear?: Revaluating the Threat of Modern Genetics', *Emory Law Journal* **39**, 678–82; 684–96.

Glass, B.H.: 1965, *Science and Ethical Values*. University of North Carolina Press, Chapel Hill, NC.

McKusick, V.: 1992, 'The Human Genome Project: Plans, Status, and Applications in Biology and Medicine'? In: Annas, G.J. and Elias, S. (eds.), *Gene Mapping: Using Law and Ethics as Guides*, Oxford University Press, NY, pp. 18–42.

Plato: *Republic*, III (410) IV, pp. 456–461.

Proctor R.: 1992, 'Genomics and Eugenics: How Fair is the Comparison?', In: Annas, G.J. and Elias, S. (eds.), *Gene Mapping: Using Law and Ethics as Guides*, Oxford University Press, NY, pp. 75–93.

Ramsey, P.: 1970, *Fabricated Man: The Ethics of Genetic Control*, Yale University Press, New Haven, CT.

Sperry, R.W.: 1974, 'Science and the Problem of Values', *Zygon* **9**, 7–21.

U.S. National Institutes of Health: 1985, 'Points to consider in the Design and Submission of Human Somatic Cell Gene Therapy', *Federal Register* **50** (160), pp. 33463–33467.

Walter, L.: 1991, 'Ethical Issues in Human Therapy', *Journal of Clinical Ethics* **2**, 267–74.

PART II

THE CONCEPT OF HUMAN NATURE:
THEOLOGICAL AND SECULAR PERSPECTIVES

KIDO INOUE

THE ZEN WORLD AND THE MENTAL GENES

I. INTRODUCTION

Nowadays we are conscious of the fact that besides responsible choices exercized through our freedom, the kind of genes we inherit play an essential role in shaping history and creating cultures. The genetic transmission of certain features from one generation to another determine the quality of life and the survival of the species.

Genes are carriers of hereditary traits and are contained in filaments of DNA that are present in the living cells of all organisms. They encode the information for building the body and making it perform its vital activities. Scientific advances into the structure and composition of our genetic make-up holds out immense prospects for gene therapy. Screening for most genetic risks will eventually be possible through analysis of chromosomal aberrations. But this same optimism resulting from our growing knowledge of human genetics is fraught with all manner of genuine dangers and has a mind-boggling side to it. One wonders towards what kind of society and what new planetary equilibrium we head!

In this reality humanity is tested in the choices it makes for itself. And in this reality we are given the opportunity to determine the fate of future generations through wise and loving choices, or through evil deeds. An individual's fate is determined by the long thread of past generations, shaping the direction of a nation and the earth itself. This is because each individual's will is not commanded by an awareness and sense of value of one's own, but by that awareness and those values passed on to the individual by the preceding generations. In this sense, an individual comprises a whole, and the whole forms each individual's awareness: when brought together, it gains force and shapes an era. Power such as this has brought changes to the world in every generation.

We are now obliged to construct a sound future in the midst of the current critical situation on earth. Scientific knowledge per se is neutral; it is neither harmful nor beneficial. It is the foundation for scientific progress that calls for wise and prudent choices that respect human dignity

E. Agius and S. Busuttil (eds.), Germ-Line Intervention and our Responsibilities to Future Generations, 21–26.
© 1998 *Kluwer Academic Publishers. Printed in Great Britain.*

and promote the quality of life without stifling the spirit. As a Buddhist priest, I would like to focus my contribution on the spiritual dimension of the individual, and address the issue of genetic engineering and our responsibilities to future generations from a Zen perspective.

II. BASIC SPIRITUAL STRUCTURE
BASED UPON THE INSTINCT TO SUSTAIN LIFE

Our genes originated with a dramatic movement to sustain, reproduce, and proliferate our life. This is common to the genes of all living things. It is a great wonder that the diversity of life and the sustainability and proliferation of species require the direct contact and relations between human beings and other living organisms. All the living phenomena are a drama of creation and evolution produced by the world of nature which has been under the control of genes so that it can evolve in an orderly manner, obeying certain laws of nature. The precious balance in the natural world is maintained through the close relationship among various species.

Our genes are a drama and a story according to which living phenomena are performed. They are based upon the desire and persistence to survive as a species. The genes themselves rest on permanence and, therefore, they generate younger generation before they perish. This is the source of sexual desire and its emotional energy drives us to sexual activities. As part of the animal world, human beings need sexual activities in order to maintain their species. At the same time, were it not for sexual desire, we would not be animal-like yet capable of self-respect, shame, and self-control.

In order to nourish our body and satisfy our hunger, we are driven by our appetite to eat. No matter how great the compunction about killing sophisticated and complex animals is, we dare to eat other living organism to survive. Therefore, in extreme situations, our instinct for survival comes into full play, and we tend to give priority to self-conservation over everything else. Once all endurance has collapsed, self-indulgence becomes our rule of life. Pushed to an extreme, this primitive spirit of self-absolutization and negation of others might lead to the violent phenomenon of killing one's parents, brothers, sisters, and friends. Through our genes, (inherited from our ancestors) these brute features enable us to survive and propagate the life of the species. Somehow genes are a property of the species and not of the individual. They do not die with the individual, but survive in other living members of the species. Genes have a universal character.

We also managed to perpetuate our species by defending ourselves against enemies, emergencies, and even the least harmful events. It is thanks to this defensive instinct that we have survived. Our instinct, still inherent in ourselves and incorporated in our spiritual emotions, works to form the human pattern of light and darkness. This pattern includes alertness, grudge, hatred, anger, fear, defence, aggression, pretence, suspicion and murderous design. From this derives our desire to protect our kin. From this crops up conflict. It is exactly this instinctive pattern that gives birth to religions, philosophies, and arts. It is the element and source of emotional energy.

This instinct to preserve our life and perpetuate our species also moves us to bring forth healthy children, and to develop love and affection. In case our kin are faced with danger, the affection toward them drives us to guard them and ourselves against enemies. Primitive love comprises the aggressive and defensive instinct to fight enemies. Once this aggressive instinct comes into play, our spirituality is weakened and we start thinking in terms of extremist dichotomy: enemies and friends, profits and losses, survival and death. Such dualistic thinking will influence all our behaviour and perception of the world around us. When we fall into such a narrow dichotomy and confrontation, every attack on us will be met by revenge. The energy and destructive power springing forth from such emotions cannot be tamed by our acquired knowledge and thinking. No culture or ideological manipulation can ever banish completely the irruption of these rebellious emotions within us. Various racial conflicts appear to be economic struggles, but in fact, they stem from our contentious instinct which has its origin in our genes. From such a view point, there is nothing that differentiates us from other creatures.

III. ENVIRONMENT AND CHARACTER BUILDING

We hope to live a significant life in peace, in spite of the limited and transitory life on earth where we all have to face death. Actually, our orientation toward a safe and comfortable life, through the flight from hazard, uneasiness and pains, is also a living phenomenon. Therefore, it is only natural for us to pursue peace. No sophisticated theory is necessary to prove this and any opposition to our deep desire for peace will prove to be groundless.

Our inherent desire for peace, however, is hampered and stifled by various kinds of egoism – individual, racial, national, human, contemporary, civilizational, ideological, religious and authoritative – and the consequence is the continued production of weapons, massacre, and nuclear tests all over the world, all of which are alien to peace. Why?

The origin of these deplorable phenomena is our *karma*, a sad existence dominated by the uncivilized mind of selfish and egocentric self-absolutization and negation of others. It is also the result of suspicion and doubt stemming from the primitive instinct to fight enemies. Our intelligence and culture are tossed around in the hands of animal-like instincts, thus bringing about every kind of folly.

Deeply concerned about the ways and means that will help us to overcome this egocentric tendency, we have to contemplate how we can contribute for the building and enhancement of a sound spirit in the next generation. Genuine wisdom calls for elimination of our instinct to fight enemies in an attempt to elevate our inherent affection to our family and kin to the absolute love. This is the way to overcome the ego.

We have to be careful not to allow human spirituality to be negatively influenced by elements in our environment that tend to stress our egocentricity. Otherwise we might activate the corresponding potential genes to such an extent that they become part and parcel of our spiritual structure. Especially in early childhood years, when the social character has not yet been developed, exposure to unnatural phenomena such as uneasiness, fear, sadness, vexation, anger, hunger and loneliness, must be avoided. Frequent contact with such uncomfortable phenomena would stimulate the child's instinct to combat enemies, including uneasiness and alertness, and reinforce the egoistic traits of suspicion and resentment to become incorporated in its process of spiritual development. The child, deprived of caring and loving emotions, would then be unable to think otherwise and respond positively in a way that promotes peaceful coexistence and harmony. If we want to promote a culture of peace and cooperation, we must wage war against our inner egoistic spirit and educate children in an environment of mutual trust, meaningful sacrifice, and contemplative spirit. To cherish healthy emotions and lofty ideals, it is indispensable to facilitate the child's contact with the world of nature and let the "silent music" therein awaken the feelings of solidarity and harmonious living.

IV. ZEN AND SELF-TRANSCENDENCE

There are diverse human characters. Some people do not hesitate to kill others to satisfy their greed. Others think nothing of themselves and are willing to sacrifice themselves for truth, peace, and people of the world. Still others try to pursue higher values in an effort to establish absolute peace in themselves.

Our intelligence is the information acquired and enhanced through the surrounding environment and personal effort, and in this sense, it is a culture. A lofty spirit is also an acquired culture which can be developed and communicated to others by charismatic leaders and our own efforts to open up to the transcendent absolute self. What is indispensable to the peace and survival of humankind is education, discipline, and genuine efforts to go beyond the ego, and the transmission of sound spiritual genes from one generation to another.

Zen, with its emphasis on the discipline of the mind and spirit aims at transcending the conflicts in the inner world to establish the absolute world free from egoism by elucidating the nature of our mind. The mind itself attends to the dynamism of ideal and emotional functions such as thought, judgment, will, and sentiment. It is not concerned with things of a transitory nature, such as instinct and our egoism, but with the eternal and immortal world.

This absolute, momentary world is exactly the eternal and immortal world without before or after, past or future. Once you become aware of the world, you would give up your ego as self-consciousness and identify yourself with the universe as it is seen, heard, or perceived. Limitless in terms of time and space, the world would become universal without any "god" which is another term for ego. Love would become absolute love, namely charity, casting light on the world. With self-transcendence, our instinct to fight enemies would be replaced by a generous activity that emphasizes gentleness, compassion, love and peace. When one attains such a highly transparent spirit, those with reason would hold others with reason in awe; that is, the transcendent, absolute self would ask them whether they themselves are really trustworthy, or whether they genuinely hope for peace. Both parties would acquire the ability to know the truth and respect one another.

In this way, our primitive instinct would be overcome through self discipline. We must understand how necessary it is constantly to rise above the normal human emotions whether of lust, anger, jealousy, or

hatred and seek the higher virtues of compassion, benevolence, and love
– as well as disinterestedness. We must train as many people as possible
to become spiritual leaders whom everyone can trust, and through whose
help others could be initiated on the path of "right mindfulness" where all
worries and transitory feelings common to human nature become tran-
scended and self-possession realized. Only by encouraging mutual respect
and harmony within a community can we hope for a natural environment
where future generations will feel at ease to take upon themselves the
virtues of compassion, love, and peaceful coexistence.

Shorinkutsu Seminary for Buddhist Priesthood
Kyoto Forum
Japan

MORAL REASONING IN BIOETHICS AND POSTERITY

In the context of normative ethics, two main theories in moral analysis that have provided a framework for evaluating moral judgments are the deontological and the teleological perspectives. These two approaches are usually distinguished on the basis of two opposing views: while teleological argumentation as commonly understood affirms that we have to evaluate each action primarily by analyzing its consequences and the end to which it aims, deontological argumentation maintains that actions are inherently right or wrong with little attention to their consequences. These opposing ethical theories underlie most of the disputes in normative ethics. Accordingly, some knowledge of these theories is indispensable for moral evaluation in bioethical issues because the extant literature continues to draw on methods and conclusions germane to these theories.[1]

Unfortunately, a sharp distinction between teleological and deontological ethics, as these terms have been historically understood, has recently become confused. In reading current bioethical literature, it is important to steer clear of these confusions, which come from three main sources. The first mistake is to identify teleological reasoning with consequentialism, and to say that teleologists determine whether an action is right or wrong only by its consequences, while deontologists maintain it can be right or wrong no matter what its consequences are. The second is to identify deontological reasoning with the acceptance of exceptionless norms, i.e., rules whose violation no circumstances can justify, and teleological reasoning with the view that under certain circumstances any norm can be broken. The third is to speak of mixed systems which are in some respects deontological and in others teleological.

The first of these confusions arises from a failure to notice that not all types of teleological ethics need be consequentialist, nor need deontological ethics ignore the moral significance of the consequences of human actions. The second confusion arises from the failure to notice that exceptionless norms can sometimes be derived in both teleological and deontological systems, and in both systems circumstances can modify the morality of an action. The third results from the failure to see that although these two methodologies are not necessarily exclusive, it is impor-

E. Agius and S. Busuttil (eds.), Germ-Line Intervention and our Responsibilities to Future Generations, 27–33.

tant to decide which is primary and which secondary and instrumental, and not merely to mix them in an eclectic manner which can only lead to paradoxes and misunderstandings.

Plurality of perceptions establishes the moral perspective as much as two eyes are fundamental to an optical perspective. A healthy moral dialectic never baptizes one ethical theory as the one and only perfect guide to be held dogmatically as binding without the delving thrust of critical argument. In our approach to moral argumentation in bioethics and the future of humankind, we shall attend to the insightful perspective of the above-mentioned theories, and appropriate what is relevant in their compelling arguments, while indicating any shortcomings in either theory. In brief, we will try:

1. to see how sometimes the deontological perspective in the field of genetics, with its tendency to disregard consequences, can hinder us from reaching positive and beneficial effects;
2. to see how the teleological perspective has to refer its position to the consequences of actions, and to the values underpinning our choices and actions; and
3. to be constantly aware of all possible consequences and try to evaluate their impact on present as well as future generations.

It is very unlikely that the future will annihilate either of the two approaches in normative ethics. But we can surely hope that, through a better understanding of the specific characteristics of both moral arguments we will arrive at a more fruitful synthesis that fully respects the dignity of the human person.

I. FUTURE GENERATIONS FROM A TELEOLOGICAL PERSPECTIVE

When normative ethics calls into question the safeguarding of future generations and attempts to consider the long-term consequences of our present actions and decisions, it is mainly operating within a teleological perspective. In fact, a correct application of the teleological theory would entail considering:

1. immediate as well as long-term consequences of our actions;
2. all persons, whether living or not yet living; therefore, both consequences for living people and future generations must be taken into account; and
3. the consequences of our actions on the whole ecosystem.

In considering these three criteria an axiological hierarchy has to be taken into account where all values are to be submitted to the highest value, namely the well-being and dignity of the human person. In conflicting situations the overriding value is always the inherent dignity of the human person, a principle rooted in all cultural traditions as far as I can tell. Protection of human dignity and worth puts discreet limits on tampering with the reservoir of potentials inherent in the "gene pool."

The well-being of future generations is presumed in the teleological perspective because we cannot assess the morality of human actions unless future consequences, at least those foreseen, are seriously considered. Moreover, the teleologist's reference to the principle of impartiality implies that all foreseeable consequences of today's actions on future generations must be carefully studied. On the basis of this principle we have to attribute the same value to each human person, whatever the geographical context and the historical period of one's existence is or may be.

II. FUTURE GENERATIONS FROM A DEONTOLOGICAL PERSPECTIVE

When the deontologist in normative ethics follows specifically his way of argumentation, the consequences of his action, whether immediate or long-term, are usually not taken into consideration seriously. Typical expressions of his fundamental principle of argument are:

1. no matter what the consequences, we never have the right or the permission to act so; and
2. it will be always wrong to do this or that, because it is against nature.

The deontologist thinks that some actions are to be considered "against nature" and therefore they are morally wrong, because they interfere in the physio-biological process of human reality. In this way, they are changing the fundamental functions and are disrespecting the finality of nature. Thus, for the deontologist, actions against nature (*contra naturam*) are ruled out without taking into account their possible consequences, even though they may be beneficial for contemporary human beings or future generations.

On the basis of this argument, it is quite obvious why issues related to future generations hardly find any place on the deontologist's agenda. In principle, the deontologist does not refer to the consequences of a forbidden action in evaluating and formulating a moral judgment. As a result, the issue of how germ-line intervention will influence future generations barely finds any resonance within this stream of moral reasoning.

III. DEONTOLOGICAL ARGUMENTATION IN BIOETHICS

Genetics is one of the specific areas of applied ethics in which the deontologist is obliged to apply the mode of reasoning that characterizes the deontological approach. Moreover in the field of bioethics the deontologist could employ the following two arguments which characterize his theory:

1. He should invoke the argument of *contra naturam* because each intervention in the genetic structure of human and non-human beings must be evaluated according to the following principle: it is somehow wrong to interfere with the fundamental process of life, and this prohibition cannot be overcome by the highly ambiguous notion of "diseased" genes.

Biology for the deontologist is equivalent to nature: the two words are synonymous and the two realities are one and the same thing. Nature cannot be reached by human intervention. In other words, each human intervention which changes nature, whether positively or negatively, must always be considered *contra naturam* and hence is morally unacceptable. If the deontologist does not admit human intervention in the physio-biological process of reproduction, because such interventions are *contra naturam*, how can he or she morally admit the intervention in the same structure of biological reality, namely life as such?

2. The deontologist must consider existing constraints, namely the prohibition to intervene in the most fundamental and characteristic structure of life. Human life, in fact, for the deontologist is and remains always untouchable, not only at the end of life, but also at the moment of its conception and development.

For deontologists, new genetic discoveries, although beneficial, are irrelevant because the consideration of consequences is not part of the framework of their moral reasoning. If a deontologist excludes the consideration of consequences, even in the case in which a *contra naturam* or a forbidden action has a positive consequence, he shall not change his moral judgment; also in the case of a negative consequence he will not stress the negativity of an already formulated judgment.

IV. TELEOLOGICAL ARGUMENTATION IN GENETICS

In bioethics, as well as in other contexts, the teleologist takes into consideration the consequences of actions as the starting point for the moral evaluation of human intervention. In fact, he excludes no consequences whatsoever at the beginning of the normative process. He does not consider an action as *contra naturam*, nor that he lacks divine permission to perform such actions. Instead, he feels responsible to probe into all possible arguments and take a long-term view of possible consequences before formulating a moral judgment.

The impossibility to foresee all possible consequences of an action raises a major difficulty for the teleologist, especially in genetic engineering. It will not be always possible to imagine what the results of the genetic intervention of today, which is still at an experimental stage, could have on future generations. And it will be much more difficult to foresee how we will stop the reactions already primed. However, that does not leave us helpless. What we cannot see directly or deduce, we can somehow model, construct, or imagine. Morality, precisely because it respects the different cultures and freedom of the human person, can never be reduced to a neat universal system of obligations. Rather, by distilling past experience and noticing the congruence of insight from all possible sources, it aims to provide fruitful directions towards a better future that benefits the whole of humankind and creation.

Accordingly, the teleologist, similar to the deontologist, should not take an *a priori* moral standing in relation to genetic engineering. But he should attempt to foresee all possible consequences of an action and to reach a moral certitude about the benefits of the consequences, the nonexistence of negative consequences or, at least, about the preferability of the positive consequences over the negative ones. Certainly we have divergent opinions among teleologists about the normative results, precisely because among the teleologists, as well as among the deontologists, we find different opinions in the application of the evaluation process and what is to be regarded as a value. Thus one can easily understand that the ethical normative discussion among teleologists is an ongoing revision process that is constantly refined by incoming information and the evaluation of changes that each intervention entails.

V. CONCLUSION

To conclude these reflections, we may say that even within the discussion about future generations and genetic engineering, it is very important to be aware of the complex problematic emerging from the different arguments and of the seemingly opposite ways of proceeding in the normative context. Especially when dealing with complex and delicate problems like the ones raised by germ-line intervention, which may have positive or negative consequences for present and future generations, it seems absolutely necessary to ensure the correct application of the logical rules of normative ethics while encouraging the mental habit of penetrating deeply into the moral dialectic that constantly probes new theories before making up its mind. The knowledge of the respective argumentative differences will permit not only the overcoming of emotivism that easily yields to the flimsiest of whims, but will lead to the development of a healthy bioethical dialogue that takes full cognizance of scientific advances and constantly channels new beneficial discoveries to the service and promotion of human dignity.

A proper decision must be the result of the fullest possible information. One is always under the obligation of refining one's moral sensibility, of learning more about the possible repercussions of one's actions, of delving deeper into the ultimate significance of each choice made and then pursued. For example, what does it mean to talk of a "common genetic heritage" that belongs to all humanity? Does this concept provide adequate moral grounding for banning any genetic intervention? Or should the concept of a "common genetic heritage" lead us to the conclusion that any genetic intervention is forbidden if it goes against the given natural genetic make-up?

A brave new world is no longer just fiction. We have the duty, as the present generation, to assume all our moral responsibilities and overcome a certain false shyness in venturing boldly but prudently into the promising paths of genetic engineering. This way we would be making a wise and loving contribution to uphold the moral character of humankind and strengthen the moral fiber that unites past, present, and future generations.

Istituto Siciliano di Bioetica
Acireale, Italy

NOTE

1 Perhaps it would be helpful to keep in mind J. Rawls's distinction between deontological and teleological theories, which has become influential in recent years. It concerns the relation between the right and the good. A teleological theory defines the good independently from the right, and the right is then defined as that which maximizes the good. Deontological theories either do not specify the good independently from the right, or do not interpret the right as maximizing the good.

KEVIN WM. WILDES

REDESIGNING THE HUMAN GENOME: ARE THERE CONSTRAINTS FROM NATURE?

It is obvious that for many people and organizations germ-line gene therapy is a taboo. In a survey of national, scientific, and bioethical commissions and bodies, Maurice de Wachter concludes: "If one assumes that such reports, recommendation and guidelines reflect society's opinion on the matter, we must conclude that the rather general approval which somatic cell gene therapy receives does not exist regarding germ-line therapy" (de Wachter, p. 169). One can ask: Why the difference?

De Wachter argues that there are three different, somewhat overlapping, objections that are made against germ-line therapy that are not raised against somatic cell therapy. First, there is the objection that future persons have a right to a genetic inheritance. This objection is exemplified by the statement of the Council of Europe on gene therapy (Council of Europe, p. 328). The second objection is "borrowed for the social order." This objection is concerned with a fear of abusive state and social authority. The objection voices a fear that coercive state authority will be used to compel germ-line testing, therapy, and the like. The third argument is based on an appeal to nature. The argument is that modifying the genetic structure of the yet to be conceived is "playing God." This objection seems to embody a powerful moral constraint for many people. The view makes two crucial assumptions. First that there is an identifiable human nature and second that this nature is morally normative. It is to the third objection that I will devote special attention in this essay. The essay will argue that the appeal to nature does not support moral constraints for a secular, morally pluralistic society. However before turning to the argument from nature we need to investigate what are the possible ways to construct secular moral constraints. The first section of this paper will investigate the possible ways to ground moral constraints. The failure of these constraints help us understand why the appeal to nature fails. In the second section I will turn specifically to the argument from nature. Finally, in the third section I will argue that only consent – or lack of consent – can serve as a basis for general secular ethics.

E. Agius and S. Busuttil (eds.), Germ-Line Intervention and our Responsibilities to Future Generations, 35–49.
© 1998 *Kluwer Academic Publishers. Printed in Great Britain.*

I. SECULAR ETHICS AND MORAL CONSTRAINTS[1]

Four Ways to Resolve a Controversy

Anyone who turns to the problems of bioethics finds the shadow of a substantial skepticism regarding the intellectual justifiability of moral claims. Moral controversies appear beyond resolution. They are passionate, but, outside of the context of a particular set of moral assumptions and premises, they do not appear resolvable. Any attempt to take bioethics seriously must begin with specifying the conditions under bioethical controversies that can be resolved with intellectual justification. Controversies about proper conduct can be resolved in only four ways – on a basis of (1) force, (2) conversion of one party to the other's point of view, (3) sound rational argument, and (4) common agreement.

Force by itself carries no moral authority, for by itself force does not give a rational justification for its use. Whether it be the force of a mob, or the state, or the will of the majority, force, by itself, fails to answer questions about why one ought to act in a certain fashion or submit to particular constraints. Using force, even the force of a democratic state, to prohibit an action, does not, by itself, provide a moral justification. It is simply the use of force.

Ethical controversies can be resolved by conversion. That is, individuals can come to see, believe, and then become part of a particular moral community such that the controversy evanesces. Indeed, this has been the traditional hope of the Christian West: conversion of all to a single, authoritative moral viewpoint, available through revelation and reason (Pius XII). But the very history of the Christian West illustrates the difficulties with this hope. It does not appear likely that all will convert to one moral understanding.

Against these frustrated aspirations of faith it is natural still to hope that reason will provide justification for a universal moral account or narrative. Indeed, the appeal to rationality seems at first especially promising. If one is able to provide a definitive rational account of a moral issue, this should resolve all the rational questions advanced by rational individuals. In short, rational individuals could not protest a definitive rational answer to a rational question without declaring their irrationality. The appeal to rationality thus comes with great promise.

This third possibility for resolving moral controversies has been a central part of the Western culture since Plato and Socrates. It has deep historical roots in the natural law tradition of the West. Roman law, while acknowledging the practices and customs of different cultures was shaped, under the influences of Cicero, Gaius, Ulpianus and Justinian, by a belief in the *jus naturale* which was known to all animals and the *jus gentium* which embodies what reason commands of any rational agent (Justin, II, 1,2). After the collapse of the Middle Ages' synthesis of faith and reason, the modern age attempted to provide rational justifications for the Judeo-Christian morality without faith in the Judeo-Christian God. Indeed, the hope of developing a rational, content-full[2] moral theory became the hallmark of the Enlightenment (MacIntyre, chapters 4 & 5).

The fundamental conceptual difficulty for the project of resolving moral controversies on the basis of rational argument is that one needs a rational standard. Such standards have been sought in (1) the very content of ethical claims, or in intuitions, as self-evidently right; (2) in the consequences of actions; (3) in the idea of an unbiased choice made by an ideal rational observer or group of rational contractors; (4) in the idea of rational moral choice itself; or (5) in the nature of reality. None of these strategies can, however, succeed because there is no way uncontroversially to select or discover the right or true moral content in reason, in intuitions, in consequences, or in the world.

The appeal to intuitions fails, because for any one intuition advanced, a contrary one can be advanced with equal ease. The same can be said with regard to systems of intuitions. What for one individual will appear to be a corrupt or deviant moral intuition can for another appear correct, wholesome, and self-evident. For example, some men and women will have a moral intuition that it is morally appropriate to abort a Down's syndrome fetus while for others such an intuition will be immoral.

The appeal to consequences faces the problem of how to assess and evaluate sets of consequences. A consequentialist will have to build in some moral view in order to evaluate possible outcomes and to know which outcomes are more important and which preferences are to be given priority. For example, one might agree that the proper goals of political life include liberty, equality, prosperity, and security. Though people may be in agreement with regard to these major goals, consequences cannot be assessed in terms of these until a ranking has been established. Depending on whether liberty or equality is given prior ranking, one will come to quite different judgments regarding the correct structure of a good society. Consequentialist accounts are no better advantaged than intuitionist accounts with regard to being able to demonstrate which set of outcomes is to be preferred since such a judgment requires an authoritative means of ranking benefits and harms. We are left in a position that one way of weighing consequences can always be countered by another way of weighing consequences with no way to judge between them except by appeal to our own moral sense.

Others have attempted to develop content-full, authoritative moral conclusions by employing some variety of hypothetical-choice theory. In such theories the ideal observer or decider, needs to be informed of the various possible choices and be impartial in weighing everyone's interests and siding with none of the parties involved. But if the observer is truly impartial how will decisions be made?

Impartial observers can make choices when criteria are agreed upon in advance. For example, the judges in an ice skating contest can judge because there is criteria established before the contest. However, the criteria are exactly what are disputed in moral discussions. There are different accounts of what constitutes a good outcome (e.g., long life or painless death). If the observer is so impartial or dispassionate so as not to favor certain outcomes over others the observer will not be able to make a choice. If the observer can choose then the criteria must have been built into the process. Despite the guise of impartiality, proponents of hypothetical-choice theories must then build into the observer some particular moral sense or thin theory of the good in the order of choice. One can see this in John Rawls' *A Theory of Justice*. By imposing particular constraints on his hypothetical contractors, Rawls presupposes that his contractors have a particular moral sense. They must (1) rank liberty more highly than other societal goods; (2) be risk aversive; (3) not be moved by envy; and (4) be heads of families. Again the problem is that the description of the contractors is one that presupposes a particular moral point of view. But one is given no independent reason to accept one particular view of the contractors over any other.

Attempts to discover a concrete view of the good life, or justice through the analysis of the concepts themselves suffer the same difficulty as hypothetical-choice theories. One must know, in advance, which sense of rationality, neutrality, or impartiality to use in choosing among different accounts of the good life, justice, or morality.

There is no content-full moral vision which is not itself already a particular moral vision. One cannot choose among alternative moral visions or thin theories of the good without already appealing to a moral vision or thin theory of the good. Any ethical theory that addresses particular moral questions must presuppose a particular set of moral commitments and view of moral justification. Without such commitments moral language can take on so many meanings as to become meaningless. Bioethics illustrates and confronts this problem because it must address particular moral questions in a systematic fashion.

Bioethics and the Appeal to Reason

In the last thirty years, with the emergence of secular bioethics, many of the philosophical attempts to develop a content-full morality from reason have been replayed in bioethics. We find appeals to consequences (Singer) and hypothetical-contracts (Veatch), rational decision makers (Daniels), and natural law. But, these projects tend to be carried out without addressing the foundational questions of moral philosophy. In so far as that is the case, they still suffer from the same foundational and conceptual questions that have plagued moral philosophy in general. They incorporate a particular moral sense to choose accounts of consequences or to discern which account of hypothetical choosers should be endorsed. But, as has already been noted, the rational project of modern moral philosophy appears to have failed.

There have been two interesting attempts to offset the theoretical dilemmas faced in appeals to rationality in order still to be able to justify particular accounts of bioethics. One well known attempt is that of Beauchamp and Childress (Beauchamp and Childress) and their appeal to mid-level principles. They acknowledge the reality of moral pluralism and the impossibility of deciding between moral theories. They try to meet the problem by developing four "mid-level" principles which persons from different moral frameworks can agree upon and use to resolve controversies. This effort has been central to the development of bioethics. However, it is freighted with difficulties (Brody; Green; Clouser and Gert; Lustig; DeGrazia).

There is the problem of determining the meaning of any of the four principles. While Beauchamp and Childress say they agree on the principle of autonomy, "autonomy" in fact has several possible meanings. Just as Kant and Mill both speak of "autonomy" while having different meanings in mind, one suspects that Beauchamp and Childress use the term in different ways. For a preference utilitarian, autonomy will be concerned with the liberty to pursue preferences. For a deontologist autonomy is not concerned with the pursuit of heteronomous desires but with the demands of reason imposing the moral law. So while the two may use the same words they really speak two different languages and must in addition mean different things. The language of bioethics goes on holiday from moral practice.

There is a second problem with this approach in that it is never clear how the principles are related one to another. Each of the principles is conceived to be *prima facie* binding. That is, they are all of equal weight. How, in conflict situations, are we then to decide rationally which principle to follow?

Both difficulties arise because Beauchamp and Childress have taken the principles out of the context of any theory. In Western philosophy principles have traditionally been a part of a comprehensive structure which begins with some first principle(s) and moves to secondary (mid-level) principles. Seeking to avoid the difficulties of theoretical accounts, Beauchamp and Childress have attempted to excise the secondary principles from any type of comprehensive structure. However, shorn of theoretical and contextual moorings the principles become incomprehensible and incoherent.

A second attempt to offset the difficulties of moral theory has been to revive casuistry (Jonsen and Toulmin) For example Jonsen and Toulmin provide an historical account of casuistry, and call for the use of casuistry in the post-modern world. However, they never develop an account of how a secular casuistry would work. Traditional casuistry was built upon particular cases and their resolutions which were paradigmatic for moral dilemmas. The difficulty with a secular casuistry is that there is no way to decide which cases are to be the paradigm cases. Again, like the appeal to middle level principles, casuistry, shorn of a moral viewpoint offers no way to select the central cases to make the machinery run. Furthermore, in the practice of casuistry the confessor had leeway in which to select the cases on which to model the case of a penitent. Even if a secular casuistry could develop a set of paradigm cases, outside of any particular theoretical and cultural framework, there is still no non-arbitrary way to choose which case should be the model for the dilemma before us outside of a particular moral or theoretical context.

Both the appeal to middle-level principles and to casuistry presuppose a moral context not available in a general secular context that is morally pluralistic. Beauchamp and Childress, on the one hand, and Johnson and Toulmin on the other have attempted to take moral strategies, which succeed within a content-full moral context, and to apply them in circumstances where a common canonical moral content is in fact not available. The general secular context is committed to no particular set of moral values or moral justification. To the degree that it is morally pluralistic there will be many different views of bioethics.

II. WHICH NATURE? WHOSE MORALITY?

In moral philosophy and bioethics there has been one other major frame of reference that has served to generate moral constraints. It is an appeal to the order of nature. Indeed, many of the arguments against germ-line therapy implicitly appeal to this framework. These are the argument de Wachter characterizes as those which judge that germ-line therapy is "playing God."

According to Munson and Davis there are usually three forms of this appeal to nature raised against germ-line therapy (Munson and Davis). The first argument views germ-line therapy as a prelude to positive eugenics. Yet it is not clear why this argument should be taken seriously. If the argument from nature constrains us from engaging in positive eugenics because positive eugenics tamper with nature then it would seem to me that we are equally constrained from ever treating disease of any kind since disease is part of nature. The assertion against positive eugenics is far more complex than it first appears. Such views usually assume that there is some objective idea of human health that enables us to distinguish therapy and enhancement. However, we know that it is notoriously difficult to establish such norms of health that are not tied to the values of a particular time and culture.

There is also a curious scientific assumption often made in these moral arguments. The human genome is described as if it were some fixed object when, in fact, it is an abstraction and not a natural object. The human gene pool has no benchmark (Juengst).

The second argument from nature is that there are unpredictable losses in germ-line therapy. However, the fear of "unforeseen disaster" is no greater with germ-line therapy than it is with somatic cell therapy or with any medical therapy. Medicine is a science based on statistical knowledge. The potential for the unforeseen disaster (or good) is always a possibility. While such concerns may justify supervision and oversight they do not, a priori, exclude germ-line therapy on the basis of potential, unknown hazards.

The third argument from nature is that germ-line therapy is a threat to our human nature. It is this objection, I think, that is at the heart of the appeal to nature. Authors like Leon Kass, or Hans Jonas see the possibility of accumulated changes over generations, that may lead to the development of beings that are "other than us." They believe that an appeal to human nature forms a basis for moral constraints on germ-line therapy. Munson and Davis argue that the sum of such changes does not warrant refusing to develop techniques for eliminating genetic disease. While I would agree with the Munson–Davis response, I think there are even stronger, more powerful objections to the appeal to a canonical view of nature. An historical review of the natural law tradition and the appeal to nature reveals the weakness of this framework as a basis for moral argument in a secular, morally pluralistic world. It is clear that there is no natural law tradition but there are natural law traditions. And therein lies the problem.

The Stoic philosophers are some of the first to appeal to nature as a basis for determining human conduct. The Stoic view centers on the assumption that the whole universe is governed by laws which both exhibit rationality and which can be discovered by human reason. Human beings, following the rest of the universe, also have essential nature which is law governed. Determining what is the essential nature is a prerequisite for determining the moral constraints.

One approach to distinguish those aspects of human life that are essential from those that are non-essential is to appeal to the ends or purposes of human life. One finds this type of teleological model in Aristotle and Thomas Aquinas. Aristotle thought that there was a defining end to human life, *eudaimonia*, which is tied to the flourishing of human nature. For this to occur one needed to be schooled in certain virtues. Aquinas brings together the Stoic principle that we ought to "follow nature" with an Aristotelian view of nature as a teleological system. For Thomas natural moral law is part of divine reason that is accessible to human intelligence (*ST*, q 91, a. 1–3). It is not to be confused with the biological or physical order which have their own laws. The precepts of the natural moral law take the form of something to be done. That is, they outline the basic human goods that ought to be pursued. The pursuit of these goods is tied to Aquinas's view of human essence and ends. The basic goods are pursued so as to effect the human. The goods are ordered so that one may achieve the end of human life (happiness or blessedness). In turn the goods give moral constraints and directions for how one ought to act.

That is, they direct us in what we should pursue and what cannot be done. We find here a very particular view of human nature that is tied to assumptions about the divine nature of all creation, particularly the human, and the divine ordering of creation.

One of the shifts that takes place in natural law thought is a move away from the theological framework of the natural law. By the seventeenth century the appeal to nature was used by the sciences as a way to talk about the natural world. The emphasis was no longer on the human participation in the divine law. Instead the emphasis was on regular, measurable realities of physical nature (Ruby). Teleology was abandoned for different forms of mechanization. The appeal to nature no longer had, necessarily, moral dimensions.

These shifts away from religious, teleological framework to mechanistic views of the world are reflected in the natural law thinking of the seventeenth and eighteenth century. Following the Protestant Reformation, Grotius and others moved away from the theological and teleological metaphysics used by Aquinas. They argued that the fundamental nature of the human being was the possession of reason. They were not concerned with developing an account of basic human goods to support a certain teleology. Rather the natural law could be understood as what reason discovers. Of course this position leads to further questions about the nature of reason.

This brief recounting is not meant to be an exhaustive overview of the natural law. The point of the overview is to illustrate a profound conceptual issue for any appeal to human nature. Any appeal to nature as the basis of moral constraints depends on how nature is understood for the constraints that will be developed. Outside the context of any particular moral framework there will be numerous ways to understand nature. Absent a common held view of human nature, we will not be able to develop content-full moral constraints.

At first glance many people will think that an appeal to nature is sufficient for moral guidance. Yet, as one investigates it becomes apparent that to understand the meaning of an appeal to nature one must situate the term in context of moral language. Outside of a particular context of a moral language terms such as 'nature' can take on multiple meanings so as to become meaningless. One can call to mind the thirty year debate in Roman Catholicism about the moral evaluation of birth control. There are those who evaluate the physical act and see artificial contraception as illicit while those who argue for a person-centered ethic see the act as

morally neutral and put the evaluative emphasis on the intention of the
agent. Yet both arguments appeal to nature.

The problem then is clear. If people agree that an appeal to nature is
paramount in our moral analysis it will depend on which understanding of
nature is brought to bear on the issue. Appeals to nature are embedded
within the context of a moral world view. Different appeals will often be
incommensurable with one another even though they appeal to nature.

Even if we could resolve the first fundamental question about the
nature of "nature" we are left with a second, perhaps, more difficult issue:
Why should nature be morally normative at all? That is, unless one begins
moral analysis by viewing nature, however defined, as morally normative,
then there is no reason as to why one should think nature as morally
normative. We are caught in a vicious circle. This objection revisits the
concerns raised by Hume's analysis of the is-ought distinction and G.E.
Moore's concern about the naturalistic fallacy. Hume argued that simply
because we can describe how something is does not mean that we can
deduce an ought from the is. Even if we could establish a common understanding of human nature, we would still face the question of why
nature should be normative morally. If one sees nature as the outcome of
random chance there will be no reason to view nature as normative.
Indeed, one may well hold the view that the natural moral imperative is to
give rational, human control of the designing process.

This circle is not simply a philosopher's game. In a morally pluralistic
world the assumptions with which one begins moral argument are significant. That is, unless men and women share the same assumptions (about
nature) then they will not follow arguments that hold that nature gives us
moral constraints.

Summary

One will not be able to resolve moral controversies simply by appealing
to the structure of reality. The difficulties here are twofold. First, in order
for the structure of reality to serve as a moral criterion nature must be
shown to be morally normative. But in the absence of some metaphysical
account of reality, it will be impossible to conclude whether the structure
of reality is accidental or morally significant apart from the concerns of
particular persons or groups of persons. This is especially the case with
regard to human nature, which appears in scientific terms to be the outcome of spontaneous mutation, selective pressures, genetic drift, con-

straints set by the laws of physics, chemistry, and biology, as well as the effects of catastrophic events. Human nature is, as such, simply a fact of reality without direct normative significance.

The second difficulty with an appeal to nature is that even if one thought that one could find moral significance in human nature, this would be possible only if one already possessed a canonical value-interpretive perspective in terms of which alternative accounts could be intersubjectively described and then subjected to a test of falsification. Even if one accepts the normativity of nature, the structure of reality is open to many descriptions and interpretations. The natural law appeal, like others, must build in some moral sense which determines which description of nature is to be normative. Like the intuitionists, and consequentialists the natural law practitioner has no rational way to demonstrate that one description of nature should trump all others.

III. CONTROVERSIES AND APPEAL TO AGREEMENT

If the attempt to establish a content-full moral vision by reason fails then the grounds for acting ethically, for respecting persons, in a secular society, are brought into question. If all becomes relative, and neither a canonical ordering of values nor moral constraints on human actions can be established, then secular morality will be one supported by force, not reason. It will have no general secular justification. Thrasymachus's position appears to have won the day (Plato, *Republic*, Bk I, 338 b). We are left with one intuition confronting another, tradition confronting tradition, ideology confronting ideology. This would mean the collapse of a central assumption of Western culture – the notion that reason can resolve moral controversies and justify a moral viewpoint. However, if one looks at the very notion of resolving moral controversies one can find a key that offers us an exit from the brink of nihilism. The key involves persons.

A fourth way to resolve ethical controversies is by agreement. If persons are interested in resolving controversies in ways other than those fundamentally dependent on force, and if God, nature and reason remain silent, then the only way to resolve controversies will be by peaceable negotiation. Or to rephrase the point, if one is interested in resolving issues peaceably without recourse to force and with moral authority, and if God is silent (i.e., one is in a secular context), and moral reasoning cannot provide canonical moral content, then moral authority can only be derived from the mutual agreement of persons. In this circumstance, one cannot discover who is a moral authority, but only who has been put in moral authority. One does not rely on reason to discover moral authority, but on will to create moral authority.

Moral authority flows from persons in their capacity to play a role in moral communities. As the source of moral authority persons are defined in terms of their capacity for both moral controversy and agreement. Thus they will need to be rational, self-conscious and possess some concept of the good. A necessary condition for the possibility of this web of moral authority is thus that persons not be used without their consent. The mutual respect of persons becomes central to the grammar of moral discourse among moral strangers. As long as resolutions to controversies are sought from sources other than those grounded in force, even if God, reason and nature are silent, there is still the possibility of a moral nexus between moral strangers. Out of this grammatical condition flows a fabric of mutual respect and the possibility for persons to create particular webs of mutual responsibilities through mutual agreement.

Appeal to this minimum notion of ethics discloses a necessary condition for the possibility of a general, secular domain of morality. It provides the basic grammar that is necessary for moral strangers to resolve moral controversies. We find something like a Kantian transcendental truth, true not of reality as it is in itself, but only of the moral reality moral strangers can share. This minimal condition recognizes that the content-full canons of moral probity will be created rather than discovered. Moreover, the necessary condition of mutual respect, (the non-use of others without their consent), is not grounded in a value given to autonomy, liberty, or persons, but is integral to the grammar of controversy resolution when God, nature and reason have failed. Appeal to mutual respect allows us to disclose a moral point of view with the fewest assumptions. To have a morality for moral strangers one needs only refrain from using them without their consent and acknowledge them as agents that can agree to or refuse to negotiate.

Appeal to mutual respect allows us to understand in general secular terms (i.e., in terms that do not depend on a particular moral vision) when force is justified and when it is not. Moral strangers negotiate peacefully and use force only in defense or in punishment for unconsented-to actions against them including the violation of agreements and covenants. Appeal to mutual respect does enables us to recognize those moral strangers who are unwilling to act ethically; that is, who insist on resolving moral controversies by the use of unjustified force. Such aggressors cannot complain in ways that are justifiable to moral strangers when they are subject to defensive or punitive force for they have set themselves outside the morality that binds moral strangers. While the morality of strangers uses language that is often associated with political theory and state craft, what is offered is not a political theory but a fundamental moral account with political implications: a language that reaches across various moral communities.

Because authority is derived from the requirement of mutual respect among persons, one finds among the salient moral institutions of the post-modern age free and informed consent, the market, and limited democracies, the last marked by robust moral exclaves created by rights to privacy. It is easy to see why circumstances are this way. If one cannot discover what patients and physicians must do in a secular context, then patients and physicians must together agree on their common undertaking. The free market provides a web of small acts of consent and cooperation where moral strangers, moved by diverse motives, collaborate together in the exchange of goods and of services. Finally, limited democracies draw upon the morality of mutual respect to provide protection from and punishment for the unconsented-to use of persons (e.g., murder, rape, and burglary), as well as to insure the enforcement of recorded contracts. In addition, they allow common endeavors with common funds, such as the creation of a basic health care system. But finally, since actual consent to particular projects must always be limited, rights of privacy must play a large role. Here it must be noted, rights to privacy are not celebrated because of any positive value assigned to such rights. Instead, they mark the limit of the plausible authority of states to intervene in the peaceable consensual actions of individuals. "Mutual respect," "negotiation," and "justified force" for their part are integral to the language which constitutes the necessary condition for a morality among strangers.

The development of a framework for moral strangers leaves us with at least two tiers of moral discourse. It allows individuals who do not share a common moral view to understand in general secular terms the moral authority of their collaboration in joint ventures (e.g. health care). This morality can provide moral justification for political institutions which hold together large scale states comprised of divergent moral communities. Like Hegel's universal class of civil servants, who serve the whole state, the appeal to mutual respect and negotiation serves the whole community of moral strangers. This universal tier, however, is content-poor. Persons will not learn about concrete moral values, goals, virtues and vices within its ambit. The content of one's moral view will be found in a second tier; the tier of particular moral communities, where one learns what are the moral virtues that are to be admired and vices that are to be disdained. Such values and virtues are not the quaint relics of our past. They are crucial to a vision of human life and of how one ought to act. But there is no one set of values which all men and women share. Persons will as a result live within their own concrete moral communities with their values and visions of life, as well as within the context of general secular morality. The different tiers of the moral life will often produce tension for the one whose life is shaped by a particular moral point of view. There will be no general secular view of how genetic therapies ought to be used. Indeed, because there are competing views of human nature there will be competing views of what constitutes health and disease that, in turn, yield different views of appropriate medical therapy. Yet each of these views appeals to nature. Since the appeals to nature will be many they will not help to justify moral constraints in a morally pluralistic, secular society. The appeals to nature return us to the moral language of permission.

Kennedy Institute of Ethics
Georgetown University
Washington D.C., U.S.A.

NOTES

[1] I am indebted to my friend and colleague, H.T. Engelhardt, Jr., for his thoughts and discussion on the issues of secularity and the limits of public moral authority. While we are "moral strangers" (in his view), or "moral acquaintances" (in my view), I have profited greatly from our conversations. I think Engelhardt goes too far in some of his thought (such as the division of the world into moral friends and moral strangers), and he is not sufficiently critical in other aspects of his thought (such as his view of "reason").

[2] I use the term 'content-full' to identify those moral views that have a commitment to a particular set of values and a ranking of those values. Two views could share a commitment to the same values (e.g., liberty and solidarity) but rank them in different orderings. In so doing you may often get very different views about what is the morally appropriate action.

BIBLIOGRAPHY

Aquinas, T.: 1981, *Summa Theologica*, Christian Classics, Westminster, MD.

Beauchamp, T. and Childress, J.: 1994, *Principles of Biomedical Ethics*, 4th ed., Oxford University Press, New York, NY.

Brody, B.: 1990, 'Quality of Scholarship in Bioethics', *Journal of Medicine and Philosophy* **15**(2), 161–178.

Clouser, K.D. and Gert, B.: 1990, 'A Critique of Principlism', *Journal of Medicine and Philosophy* **15**(2), 219–236.

Council of Europe: 1982, Recommendation 934, in *Human Gene Therapy* **2**, pp. 327–328.

Daniels, N.: 1985, *Just Health Care*, Cambridge University Press, Cambridge, MA.

De Grazia, D.: 1992, 'Moving Forward in Bioethical Theory: Theories, Cases and Specified Principlism', *Journal of Medicine and Philosophy* **17**, 511–39.

De Wachter, M.A.M.: 1993, 'Ethical Aspects of Human Germ-Line Therapy', *Bioethics* **7**, 166–177.

Green, R.M. : 1990, 'Method in Bioethics', *Journal of Medicine and Philosophy* **15**(2), 179–197.

Grotius, H.: 1925, *On The Law of War and Peace*, trans. F. W. Kelsey, Clarendon Press, Oxford, UK.

Juengst, E.: 1997, 'Should We Treat the Human Germ-Line as a Global Human Resource?', in this volume, pp. 000–000.

Justinian: 1970, *The Institutes of Justinian*. Greenwood Press, Westport, CT.

Lustig, A.B.: 1993, 'Principles: A critique of the Critique', *Journal of Medicine and Philosophy* **17**, 487–510.

MacIntyre, A.: 1981, *After Virtue*, University of Notre Dame Press, Notre Dame, IN.

Munson, R. and Davis, L.H.: 1992, 'Germ-Line Gene Therapy and the Medical Imperative', *Kennedy Institute of Ethics Journal* **2**, 137–158.

Parliamentary Assembly of the Council of Europe: 1991, Recommendation 934 (1982) 'On Genetic Engineering', *Human Gene Therapy* **2**, pp. 327–328.

Pius XII, *Summi pontificatus*, October 20, 1939.

Rawls, J.: 1971, *A Theory of Justice*, Belknap Press, Cambridge, MA.

Ruby, J.: 1986, 'The Origins of Scientific Law', *Journal of History of Ideas* **47**, 341–359.

Singer, P.: 1995, *Practical Ethics*, Cambridge University Press, Cambridge, UK.

Veatch, R.: 1981, *A Theory of Medical Ethics*, Basic Books, New York, NY.

H. TRISTRAM ENGELHARDT, JR.

HUMAN NATURE GENETICALLY RE-ENGINEERED: MORAL RESPONSIBILITIES TO FUTURE GENERATIONS

I. INTRODUCTION:
SEARCHING FOR AN ANSWER THAT CANNOT BE FOUND

The prospect of human genetic germ-line engineering raises questions regarding the propriety of altering the human genome. It raises questions as well regarding the ways in which one might understand responsibilities to the future generations who will experience the result of such alterations. This essay explores the difficulty of disclosing content-full obligations regarding genetic germ-line engineering. Instead, as this essay shows, one is at best guided by general canons of prudence and caution. There are no absolute bars, indeed, no special proscriptions in general secular morality, against the use of germ-line genetic engineering or against the project of enhancing the human genome. In particular, this essay will argue that point 8 of the "World Declaration on Our Responsibilities Towards Future Generations," namely, that one has a responsibility "to ensure the transmission of the unique human genetic inheritance, free from any engineered alterations" (Agius 1994, p. 310), is indefensible, at least in general secular terms. In deed, if one endorses the first principle, namely, that one ensure the existence of future generations, this first principle will plausibly entail altering the human genome (Agius 1994, p. 308). Not only can the proscription against altering the human genome not be defended as a general secular moral constraint, indeed, the very opposite appears plausible. There are many conceivable circumstances under which one would find oneself obliged to alter the human genome.

In addressing the difficulties of establishing special content-full moral responsibilities to future generations with reference to interventions in the human genome, it will be shown that one cannot provide canonical content for such purported responsibilities. Moreover, given the difficulty of specifying with certainty the persons to whom one has responsibility, and given the possibility that there may not in the long run even be future generations, there is a question of the governance of most of the usual

E. Agius and S. Busuttil (eds.), Germ-Line Intervention and our Responsibilities to Future Generations, 51–63.

obligations one has to actual persons. This essay concludes that not only does one lack an absolute obligation to avoid genetic germ-line engineering or enhancement of the human genome by this technology, so that under the appropriate conditions such interventions would be licit in general secular terms, in addition there is no one single vision of the goals that should guide such interventions. One is free to act, and free to act in many different ways. At most, in approaching these difficult moral questions one is guided by three contentless moral principles: first, avoid malevolent acts against future generations, second, do not undertake changes in the human genome that one has good grounds for knowing the recipients will find unacceptable, and third, act prudently so as not to cause more harm than benefit. In each case, any content (e.g., for comparing benefits and harms) must be provided by those considering the interventions, not those who will be subject to them.

Many hunger for more guidance and more substantive moral principles. Indeed, many hunger for the moral authority in this area to impose one particular moral vision on all. Neither the guidance nor the moral authority are to be found within the compass of secular philosophical reflection. Such hunger likely reflects a desire for the substance and guidance available within religious faith. However, the history of this century provides ample grounds for concern regarding the pursuit of transcendent truth in immanent social policies. There is much danger, indeed immorality, in seeking to impose one particular secular moral philosophical vision, or ideology on all as a surrogate for a common religious faith. Since secular morals allow considerable liberty in framing visions of the good and in pursuing such visions, one should be concerned with the establishment of any international guardian of the genome or, indeed, of the environment that seeks to impose a particular content-full moral vision, ideology, or understanding of ecological rectitude. Such would be the secular equivalent of establishing a particular religion and imposing it upon the entire world. As always, one must be concerned that we guard ourselves against the guardians. One should with honesty acknowledge the character of our moral predicament: One faces a serious task, namely, of using human germ-line engineering in reshaping the human genome possibly guided by a diversity of goals and constrained by only very limited moral principles.

II. PIOUS WORDS AND THE IMPLAUSIBILITY OF ABSOLUTE PROHIBITION

To establish a moral prohibition against engaging in germ-line genetic engineering in general, or in particular to convey enhanced abilities, one might hope to establish that: (1) such alterations in the human genome are intrinsically wrong because of the status of the human genome, (2) there are obligations to others or rights possessed by others that would be violated by such undertakings, or (3) such undertakings would on balance cause more harms than benefits. In short, one must show that such interventions would be contrary to either the right or the good. None of this can be established.

There is much talk concerning the inviolability of the human genome, often framed somewhat as the secular equivalent of a claim that the genome as we have it possesses a quasi-sacred standing. The difficulty lies in making out the notion of such an inviolability or quasi-sacred standing in general secular terms. After all, the human genome as we have it is the outcome of past mutations, selective pressures, genetic drift, random catastrophes, and the various constraints of natural forces and laws. The result has been to produce a human genome fairly adapted to the past environments within which humans were once found. To establish the inviolability of the human genome would require either showing that the process that produced the genome as we find it has a special moral claim on our attention, such that we should not supplement or supplant it, or that the product of this process has a sacrosanct standing. Such a claim might be appropriate, were one to consider the human genome the well-designed creation of God untouched by the consequences of the Fall. The genome as such would have a presumptive standing as appropriately designed in ways that would transcend our wisdom. However, in general secular terms the process by which the human genome has been fashioned is morally neutral and has not been optimally efficient in either maximizing the inclusive fitness of humans or in supporting the wide range of goals to which humans are committed. We can assess its appropriateness to our goals and envisage means to effect alterations that would remedy matters.

In addition, the human genome as the product of a natural process includes a set of dispositions, abilities, and functions that are at best good adaptations to a past in which humans no longer find themselves. Here one must note that the content of good adaptation depends on specifying an environment and the goals of adaptation. The environments within which the human genome was formed in great measure are no longer those of a high-technology society in which humans now live. Moreover, insofar as one would wish to apply teleological language to the fashioning of the human genome, the goal to which its structuring was directed is that of inclusive reproductive fitness. However, those goals do not have a *prima facie* claim on our attention. That is, they need not be an overriding consideration. Any account of whether one would find the human genome acceptable as it now exists, or in need of alteration, will depend on the environments in which humans are likely to live and on the goals persons would wish to realize.

Given different environments and/or goals, different interventions or alterations will be appropriate. A useful illustration of what is at stake is found in artificial means of birth control. As the Roman Catholic Church has well understood, such interventions involve an unnatural disturbance of the ways in which humans are designed to reproduce. They alter the phenotypic expression of genetically determined mechanisms for reproduction. The goal has been to make nature conform to the purposes of persons so that persons would not be constrained by nature. The use of artificial contraception involves a restructuring of human capacities so that they serve personal goals rather than necessarily those of inclusive fitness. In this relationship, persons regard their nature as a given open to criticism, judgment, and refashioning. Human nature, in particular the reproductive abilities of human, are thus objectified and altered so as better to conform with the subjective goals of persons. Outside of a particular religious understanding that would make such interventions in principle impermissible, the morality of the artificial control of reproduction is generally assessed in terms of the consent of those involved, calculations of the balances of benefits over harms, and judgements regarding the compatibility of the use of artificial contraception with the vision of appropriate reproduction embraced by those considering its use. One does not appeal to an absolute prohibition, for such cannot be understood outside of a particular moral vision.

The prospect of germ-line genetic engineering and of the altering of the human genome raise issues similar to those of artificial contraception, save for the circumstance that germ-line genetic engineering would lead to genotypic, not merely phenotypic, changes. As with artificial contraception, human nature is subject to judgement, critical reassessment, and the refashioning of persons. Whether particular interventions are judged to be appropriate or inappropriate will also depend on the consent of those involved, the balance of benefits and harms, and the particular moral visions of what is appropriate in fashioning the human future, which may drive the choices of participants. As with artificial contraception, there will not be an absolute prohibition that can be articulated in general secular terms. As we will see, there is no general secular moral basis to proscribe the refashioning of human nature so as better to meet the wishes and desires of persons.

As with artificial birth control, so, too, with germ-line genetic engineering there may be a concern that the use of such technologies may imperil the existence of future generations. After all, birth control can involve the decision of a couple not to reproduce and germ-line genetic engineering might be used to reduce human fertility. This issue can then be placed in global terms. Namely, to reconsider the alleged responsibility articulated in the "World Declaration on our Responsibilities Towards Future Generations," point 1, that there is a responsibility to ensure the existence of future generations (Agius 1994, p. 308). Since all species, planets, and galaxies and likely this universe itself are transient, it seems far fetched to hold that there should be an overriding or absolute obligation to ensure the existence of future generations. It is quite another thing to recognize an obligation not to act in ways so as to harm future generations, should they exist. But any commitment to maintaining future generations must be a particular commitment that involves particular considerations of what costs should be sustained to ensure that future generations will exist. Just as many couples have no concern for reproduction, so, too, one might very well imagine humans increasingly losing concern with ensuring that distant future generations will exist and instead focusing on maintaining the quality of life for this generation and immediate future generations (i.e., the focus would be on fewer persons and more resources for the persons who do exist). In addition, insofar as the future might pose considerable challenges to the long-term survival of the human species, meeting the challenge of ensuring the existence of distant future generations would involve significant costs. In such circumstances, attempting to ensure the existence of future generations would erode the quality of life of current and near-term generations. Absent some acknowledged transcendent moral obligation to preserve the human species, many calculations of benefits and harms would lead to the conclusion that the endeavor is not worth the trouble in that there is no one to whom persons owe a secular obligation to ensure the further survival of the human species. The inability to establish such a purported overriding goal will be further examined below.

The phenomenon of claims regarding the impermissibility of altering the human genome itself requires attention. The attempt to frame such secular philosophical moral claims may reflect an attempt to transfer to a general secular philosophical forum what were once generally understood to be religious obligations to a transcendent God. Insofar as one recognizes God, one can recognize the existence of a Person to Whom one has concrete obligations, including a concrete obligation to maintain the human species and to refrain from particular alterations of the human genome. Such a moral obligation is framed within a content-full understanding of a particular relationship to a particular God. However, once one steps away from such a particular content-full relationship and begins to speak even in general ecumenical terms, the particularity of the obligation and the content of its claim are eroded. To know under what circumstances one must do what for God requires having particular knowledge of a particular God Who makes particular demands. Matters are even more impoverished once one moves to a secular context in which a transcendent focus is lost, leaving one with only those goals that particular persons can affirm in terms of immanent considerations. Here, rather than finding a univocal content-full moral focus, as we will see, one finds a plurality of moral visions and no absolute prohibition against altering the human genome.

If one cannot in general secular terms establish an absolute proscription against altering the human genome because of a duty to God or because of a sacred standing of the genome as we find it, one might then attempt to defend such a proscription in terms of the rights of future generations to inherit an unaltered genome. But what would the force of such a right be? Consider, for example, if through the careful, cautious, and prudent development of genetic germ-line engineering one could with even greater reliability than with current forms of *in vivo* natural reproduction produce children without diabetes, with a resistance against most forms of cancer, with a resistance against environmental agents so as not to be easily disposed to developing chronic obstructive pulmonary disease, with a much decreased likelihood of developing coronary artery disease, with teeth less likely to develop caries, with a genetic resistance against AIDS and tuberculosis, and with increased intelligence and increased cardiorespiratory reserve. One would not simply have cured certain diseases. One would have enhanced the capacity of humans in a wide range of environments. In what sense would future generations have a right to be denied such benefits? What obligation does one have to future generations not to make them better off in these respects?

Perhaps what is meant by claims regarding the purported right of future generations to receive an unaltered genome is that future generations should be protected against risks, dangers, and harms possibly associated with altering the human genome, and that in fact no actual claim is being advanced regarding a right to an unaltered genome. If such is the case, the purported right of future generations to receive an unaltered genome is a circumlocution for a concern not to alter the human genome unless one has good grounds for holding that the benefits involved will outweigh the risks. The difficulty lies in determining when the risks would be such as to counterbalance the goods of engineering away diabetes, susceptibility to a wide range of cancers, etc., as well as of increasing human resistance to various currently infectious and noxious environmental agents. There is always the possibility of conjuring up certain improbable but possible high-risk outcomes associated with engaging in germ-line genetic engineering, the alteration of the human genome, and the enhancement of human capacities. However, there are also many possible future circumstances in which the ability to alter the human genome would allow humans to survive (e.g., if mutations of viruses were to lead to new and highly contagious plagues). Fantasies of possible long-range disasters from the development and use of the technology to alter the genome and enhance human capacities can be easily outbalanced with the prospect that the same interventions will convey crucial advantages.

The purported obligation not to alter the human genome turns out at best to be an obligation not to act imprudently or in ways that are likely to cause more harm than benefit. Moreover, insofar as significant benefits can be achieved and harms are less in comparison, and insofar as one considers persons obliged to pursue the good, it will be plausible to assert that there is an obligation, all else being equal, to develop human germ-line engineering and enhance human capacities through altering the human genome. It is not possible, at least in general secular terms, to defend an absolute prohibition against enhancing human capacities through germ-line genetic engineering. Concerns about the alteration of the human genome must be put in much more modest moral terms.

III. SECULAR MORALITY:
CAN AN ETHIC OF RESPONSIBILITY TO REGARD FUTURE
GENERATIONS HAVE CONTENT?

In attempting to frame an ethic committed to protecting and respecting future generations, one must establish why any one particular such ethic should have governance in preference to others. The modern philosophical project has traditionally addressed this challenge of establishing a secular ethic by disclosing within the very character of humans certain moral commitments, sentiments, and sympathies, such that all rational agents must understand themselves obliged to act in accord with those commitments, sentiments, and sympathies. In order successfully to bind these commitments, sentiments, and sympathies to moral agents as such, it is necessary to show that they are integral to rational behavior as such. Otherwise, moral agents could without fault choose other commitments, sentiments, and sympathies. Insofar as one could so establish a particular content, one would be able to dismiss all those who disagreed as irrational, claim the authority of reason in imposing behavior in accord with the established moral canons, and consider all coercive force used not as alien to the moral agents on whom it was visited, but as congenial in restoring them to their true rational character. To accomplish this, one needs to show the governance on all of the same basic moral premises, as well as the same rules of moral inference. In the absence of either of these two conditions, the project fails.

Such an approach would have provided the basis for a content-full secular morality directed both to actual moral agents as well as to future generations. The difficulty has proved to lie in providing a canonical content, background basic moral premises, not to mention common rules of moral inference (i.e., agreement about moral theory). For example, if one attempts to realize the project of articulating a morality directed to future generations, so as to maximize the beneficial consequences of any policies currently adopted, the problem is how to compare consequences at issue. For example, is it preferable to maintain the current range of adaptations or to be better adapted to riding horseback while swilling prodigious amounts of bourbon? One will need to select a canonical environment and specify the particular goals that are to be realized in that environment and then rank them. Depending on the kind of life one sees to be appropriate, one will assign different costs and benefits to particular outcomes.

Nor will it be possible simply to attempt to maximize the satisfaction of preferences. One will need to know how to compare impassioned versus rational preferences. More difficult yet with respect to future generations, one will need to know what discount rate one should employ, in that over the long term one cannot be sure if there will in fact be future generations. If one attempts to make decisions in terms of which policies will maximize the satisfaction of preferences, what numbers of future persons should one take into account and what preferences should one assign to those persons?

One encounters here a well-known difficulty in the acquisition of content for secular morality. In order to know how to rank or compare consequences and preferences so as to calculate a balance of benefits and harms, in order to know how hypothetical decision makers, including hypothetical contractors, should choose, in order to know what to regard as morally significant in nature, how, and why, one must already possess a thin theory of the good, a moral sense, a set of guiding moral sentiments, or the equivalent. But, of course, this is in fact what is at stake. In order to establish a particular ranking of consequences or preferences, in order to select a particular thin theory of the good or moral sense, etc., one must already be guided by a ranking of consequences or preferences, a particular thin theory of the good, or a particular moral sense, etc. One must already have made fundamental moral choices before one can establish a particular moral account. One is confronted then with either begging the question or engaging in an infinite regress. One cannot rationally discover content-full moral principles for secular morality in general, or for a secular morality directed to future generations.

These unhappy conclusions do not despoil us of a general secular morality. If one cannot derive a common morality from all turning to God in right belief and right worship, and if one cannot by rational argument either discover the content-full canonical morality or by sound rational argument identify the correct content-full morality without begging the question or engaging in an infinite regress, one can still (1) derive general secular moral authority from the concurrence of those who agree to collaborate and (2) identify the general commitment of morality, whatever its content, to avoiding malevolence and to achieving the good. The lineaments of a general secular morality are thus without any specific content in the sense of a commitment to a particular understanding of the good or of human well-being. Any actual content must be authorized. It must be created, accepted by particular communities through particular agreements. Under these circumstances, the most one can articulate as general principles to guide a secular morality directed towards future generations is that: (1) one should not act malevolently, (2) one should not act in ways in which one judges will violate the freedom of future persons, and (3) whatever one's schedule of costs and values, one should act to maximize benefits over harms.

The first principle turns on the circumstance that to act malevolently is to will contrary to morality. This proscription has no particular content. It does not inform one what one should take to be the good. It simply reminds us of a core element of being moral. The second principle rules out those acts which one knows will violate the will of future persons in the way in which one can have similar knowledge about actual persons. One might consider the example of planting a bomb to explode in a building fifty years later. One has the same grounds one would have regarding actual persons to know that the future person will not be con-senting to be blown up. However, in case of most trade-offs between different sets of risks and benefits for future persons, matters become hopelessly unclear. Clear cases can be constructed, such as engineering a genetic trait that will cause a severe, painful, and disabling disease that will express itself at age 8. Such examples are not very useful, since the goal of all likely interest in germ-line genetic engineering is quite to the contrary. Finally, the third principle, the principle of beneficence, the obligation to do the good, expresses another dimension of what it is to be moral. Since there are different understandings of the good, beneficence will have different content for different persons and communities. Still, it is likely that for many the principle of beneficence will have a content that will lead to affirming germ-line genetic engineering and the en-hancement of human abilities through altering the human genome, con-sidering that in one plausible account germ-line genetic engineering directed to enhancing human capacities will maximize the preference satisfaction of future generations.

The sparseness of these principles has important implications for framing the notion of an international guardian for future generations, the human genome, or the world environment. Any such guardian, if it is to have secular moral authority, will be obliged to act in a most general and content-less fashion, leaving moral communities free to act peaceably on their own understandings of the good, of human well-being, and of human flourishing. As in the area of religious ecumenism where those who believe in little can easily collaborate and those who have substantial belief will recognize each other as heretics and schismatics, so, too, in order to compass much the guardian must eschew content-full moral commitments, enforce only contentless moral principles, and tolerate a diversity of moral communities pursuing a diversity of visions of human well-being and flourishing. As a result, an international guardian that attempted thoroughgoingly to impose a particular all-encompassing vision of obligations to future generations, the human genome, or the environment would be totalitarian and properly the object of secular moral condemnation and resistance. The moral wrongness would lie in constraining without permission the freedom of persons peaceably with others to pursue their own vision of the good life and of human flourishing. Since this moral wrongness is derived from the violation of the cardinal source of secular moral authority, the actual authorization of actual persons, this wrongness would be considerable.

One should note that a guardian for future generations is disadvantaged in comparison with a guardian for present actual persons. The difficulty with being a future person is that one can neither concretely have a will nor have a view of the good life and human flourishing. Actual persons have both a standing within the general practice of resolving moral controversies by permission, as well as being the actual sources of actual permission. Future persons cannot be the source of actual permission. They can be considered as the objects of moral regard as if they were persons only insofar as they can be placed within the general practice of resolving secular moral controversies in ways that do not go against the practice of using actual persons only with their permission. But when it is not possible plausibly to understand what the object of their will would be, it is not possible to articulate any particular obligation. When one is only a general possibility, one only has very general moral standing and can make only very general claims. The one moral practice that can with general secular moral guidance provide strong claims of mutual respect does not provide much guidance.[1]

So, too, with appeals to the good. The only accounts of the good that can have general governance are those that concern considerations of non-malevolence and beneficence. However, these can receive authorized moral content only by being endorsed by actual moral agents making actual moral judgements, since there is no secular canonical content-full account of the good that must bind all. With actual persons, no community-specific account of the good is achievable in general secular terms, save through collaboration in permission. However, future persons cannot so collaborate. They cannot specifically bind particular persons in particular agreements regarding the nature of the good and character of human well-being. As a consequence, one can regard the human genome, future generations, and the environment only through the diverse moral lenses of different moral communities. It is they and only they who can give a content-full narrative or account of a proper regard of the human genome, future generations, and the environment. Such narratives and accounts are multiple.

IV. IS THERE A MORAL OBLIGATION TO GUARANTEE FUTURE GENERATIONS AN UNALTERED GENOME?

A reflection on the prospect of germ-line engineering and the enhancement of human capacities leads to a set of conclusions that many will find paradoxical. First, there is no basis for an absolute proscription. Second, content-full accounts of a proper regard of future generations and the use of germ-line genetic engineering are particular and therefore diverse. Third, it is plausible that on reflection many will see the virtue of such alterations. The seeds of the third conclusion are implicit in principle 1 of the World Declaration on Our Responsibilities Towards Future Generations. If one accepts principle 1 of the World Declaration on Our Responsibilities Towards Future Generations as an overriding obligation, such an obligation, given modest assumptions regarding possible future risks to mankind, entails a further obligation to alter the human genome genetically, so that one could ensure that humans are maximally adapted so as to survive well into the future. Principle 1 of the Declaration is in conflict with principle 8. This conflict is heuristic, for it shows the importance for humans of recasting the genetic inheritance so that future generations can receive an altered genome, insofar as this would serve goals persons might hold to be important. Given the possible long-range threat of new viruses and likely adverse changes in the environment, and insofar as one would wish to maximize the likelihood of the human species surviving, interest in genetically altering the human genome is likely to be substantial.

One ends where perhaps one did not expect: there is no absolute basis for condemning the alteration of the human genome, even if undertaken to enhance capacities of persons. Moreover, there is an obligation by default to tolerate numerous different approaches to caring for the human genome and respecting future generations. These considerations weigh heavily against schemes to enact thoroughgoing, all-encompassing, content-full international accords in this area. Indeed, the considerations advanced suggest that such accords would violate the fundamental secular moral rights of persons. As usual, human freedom and human moral diversity go hand in hand.

Center for Medical Ethics and Health Policy
Baylor College of Medicine
Houston, Texas, U.S.A.

NOTE

[1]   These arguments, as well as the general frame of considerations supporting this paper, have been developed at length elsewhere. See Engelhardt 1996.

BIBLIOGRAPHY

Agius, E. and S. Busuttil (eds.): 1994, *What Future for Future Generations?* Foundation for International Studies, Malta.
Engelhardt, H.T., Jr.: 1996, *The Foundations of Bioethics*, 2nd ed., Oxford University Press, New York, NY.

PART III

GENETIC INTERVENTIONS AND THE COMMON HERITAGE VIEW

PATENTING LIFE:
OUR RESPONSIBILITIES TO PRESENT AND FUTURE
GENERATIONS

Recent advances in biotechnology have ushered in a new epoch, one in which we have become the sovereign over the biological destiny of all living matter. The contemporary genetic engineering interventions in the evolutionary process of nature are substantially different from the traditional tampering with biological organisms. Previous human intervention on living organisms was restricted by built-in limits, such as the crossing of species borders and mating barriers. Presently, biotechnology by-passes such restraints altogether. After scientists unlocked the secrets of the genetic structure of DNA, sophisticated gene splicing techniques were developed which allow us to combine genetic material across natural boundaries, turning all of life, human and nonhuman alike, into manipulable chemical material. The working unit is no longer the organism but the gene which can now be exchanged between unrelated species.

With the new-found ability to manipulate the very blueprint of living organisms, we are assuming a new role. Biotechnology has brought us to the threshold of being able to fashion new life by altering various life forms that have evolved through a natural process. We can reprogram the genetic codes of living organisms to suit our cultural and economic needs. The biology of the planet can be remodeled, this time in our own image and desires to serve our needs. We are now entering the "Biotechnological Age."

During this last few years, a host of moral issues have emerged from the recent explosive growth of biotechnology. Controversial ethical questions have raised wide public concern and apprehension about the future of mankind. One of such question is the patentability and commercialisation of the genetically engineered "living products" of biotechnology. The idea of private ownership of life and dominion over living nature has aroused fear of abuse, especially with respect to possible eventual alteration of the human species. Should a form of life be a proper subject for a patent? Is it ethical to control and own life for personal or corporate gain? What long-range effects will the granting of exclusive rights over new life forms have on the future of species? What

E. Agius and S. Busuttil (eds.), Germ-Line Intervention and our Responsibilities to Future Generations, 67–83.
© 1998 *Kluwer Academic Publishers. Printed in Great Britain.*

biological and evolutionary consequences will intellectual property rights
have on the present and the far-distant future generations? What are the
possible effects of the patent law system practised presently by the northern hemisphere on the developing countries? These questions indicate
that biotechnology has brought serious matters at stake.

The main arena of the debate about the patentability of living organisms arose first in the United States. The Chakrabarty decision which
granted patent protection to a genetically modified bacterium was interpreted in 1988 to cover also an "oncomouse" which was genetically
engineered by scientists of Harvard University (U.S. Congress, 1988).[1]
The decision in the United States that engineered strains of animals are
covered by patent law has triggered a great debate in Europe. The patent
laws of most European countries specifically exclude the patenting of
plant and animal varieties because the International Conventions (Paris
1961; Strasbourg 1968) to which many European countries are party were
drawn at a time when biotechnological processes were either non-existent
or in their infancy. Adaptation of the texts of the conventions to cover
these new processes and products never took place with the result that
national interpretation of conventions diverged considerably.

In view of this lack of a common policy, the European Community set
up a European Commission to propose draft directives to approximate
national laws governing intellectual property rights for biotechnological
inventions. The draft proposal for a new European patent law for biotechnology was submitted by the Commission in October, 1988. It is
expected that the adoption of these directives by member states of the
European Community will induce greater intra-Community trade since it
will be easier for companies to operate within a single commercial environment. The draft proposal of the European Commission suggests also
the granting of patent protection on genetically-engineered plants and
animals. This proposal is made because the Community innovators in the
field of biotechnology do not want to be disadvantaged compared with
their competitors in the U.S. and Japan whose legislation provide more
protection through patentability than is currently available in Europe.

The moral debate prompted by the evolving patent law systems in
various parts of the world is raising two fundamental questions about its
double imbalancing effects, namely that between some parts of the world
and others on the one hand, and that between present and future generations on the other:

a) The appropriateness of granting ownership rights over new life forms
is being seriously questioned because of its far-reaching effects on the
future members of the human species. The new sorts of "products"
created by biotechnology and their potential threats to unborn generations challenge the traditional classifications of patent right law.
b) The present patenting systems adopted by the industrialised regions of
the world are widening the gap between the developed and developing
countries. Patent laws are facilitating a handful of multinational companies to control the pharmaceutical, chemical, agricultural and food
processing industries. These firms are protecting their own interests by
acquiring monopoly control (through patents) over the genetic resources and technology. The chances of the Third World to gain access to scientific information and rights to licence technology are being threatened.

Accordingly, biotechnological development is threatening the interests
of future generations in the more developed areas and of the present
generation in the less developed areas. It is in the interest of both that
another pattern of patent laws be adopted by mankind than that pursued
over the past years by the industrialised region of the world.

This article seeks to demonstrate that the current intellectual property
system must be altered for the regulation of biotechnological inventions
because of its potential threats to the future generations of the industrialised countries and to the present generation of the developing countries.
Numerous articles have challenged the propriety of patenting genetically
altered living organisms. But none of these articles has suggested how the
current patent system could be modified and improved. This article
reaches the conclusion that what we need is a World Patent Convention
on Biotechnological Inventions inspired by the concept of the "common
heritage of mankind."

I. OWNERSHIP OF GENES:
POTENTIAL RISKS FOR PRESENT AND FUTURE GENERATIONS

Intellectual property rights over new life forms can seriously affect future
generations in two ways. Firstly, the widespread practice of patenting
genetically engineered life forms can lead, in the long run, to the depletion of the gene pool of various species. Secondly, the patentability of
new life forms created from different species, and the interference with

the basic genetic structural components of living organisms are a serious threat to the unity of species. Moreover, the new technological revolution in life sciences is already having profound impacts on the poor in the least developed countries.

1. Loss of Genetic Diversity

The diversity of life forms, so numerous that we have not yet identified most of them, is the greatest wonder of this planet. A consensus is emerging within the scientific community that species are disappearing at an unprecedented rate. Though the major cause of biological impoverishment is the destruction of a species' habitat by some form of human activity, the patenting of genetically altered living organisms is also contributing to genetic erosion. Though plant and animal biotechnologies are still at their infancy, they are already demonstrating their enormous potential threats to biological diversity. The extinction of races and species of plants and animals is expected to increase, the more biotechnology will advance and expand, and the more modified living organisms will be patented for commercial purposes. Past experience with plants has shown that genetic erosion could really happen. Unless we take steps to safeguard the biological diversity of the earth's species, future generations will inherit an impoverished earth.

Genetic engineering is inducing farmers to use only the most efficient plants or animals of a species. The agricultural biotechnology is making crops more productive by crossing breeding lines with valuable characteristics and screening the progeny for individuals that have the desired traits. A consequence of this improvement is the narrowing of the genetic base of commercial varieties. Contemporary growers are lured by the patented genetically engineered cultivars whose individuals all look alike, taste the same and behave in exactly the same way in the field, after the harvest and in the kitchen. Moreover, reliance on particular strains of genetically engineered plants and animals is dangerous because of an unforeseen disease. Engineering with plants has shown that this can be a serious problem. In 1970, the U.S. lost half of its maize crop to a fungal disease called corn leaf blight. If several different types of corn were employed, the disease would not have been so destructive. This could also happen with livestock.

The possible loss of biological diversity is far from the entire story. A long-term and ultimately more serious repercussion could be the disruption of the course of evolution, insofar as the process will have to work with a greatly reduced pool of species and their genetic material. The loss of biological diversity restricts the possibility of change and evolution. Evolution thrives on genetic diversity. The biological diversity of the earth's species and their genetic material, apart from aesthetic value, represent an abundant stock of natural resources that serve our material welfare in more than we realize. The irreversible loss of species will deprive future generations from the experience, enjoyment and use of biological diversity. The extinction of species means the total loss of a possibility that can be enjoyed by generations yet to be born. The preservation of the richness of biodiversity will improve the long-term well-being of the human species.

2. Threat to Species-Unity

Species are regarded conceptually as a population or series of populations within which free gene flow occurs under natural conditions. Since the genetic system in each species differs, interbreeding is absent between species. Species are genetically close systems. In others words, species do not exchange genes. The natural barriers between species are now removed by biotechnology. Genetic engineering has given us the power to change the genetic characteristics of a living organism by transferring genetic information between species. Genes of human beings, animals and plants can be cultivated to create new life forms. Biotechnology has already opened the possibility to create new kinds of livestock with genetic traits distinct of its species or breed. If the future advances of biotechnology will extensively interfere with the basic structural components of living organisms, the possibilities of cross fertilization might be affected.

During the last decade, the prospect of direct application of gene splicing to cure human genetic diseases has moved forward by great strides, although great difficulties have to be overcome. Researchers are attempting to develop genetic splicing techniques for the use in somatic and germ cells. The great difference between somatic and germ-line therapy is that whereas the former affects only the person being treated, the latter involves changes that can be passed on to future generations. Thus far, all efforts towards human gene therapy has focused on somatic cells. If human somatic gene therapy is medically accepted, a powerful rationale will exist to extend gene therapy to germ-line cells as well.

Though genetic engineering techniques are already demonstrating their potential value for human well-being, they can be also used for positive eugenics in order to change the basic characteristics of human nature rather than to cure chromosomal disorders. One of the greatest threats of genetic engineering is the possible disruption of the stable unity of the human species who is the only survival of a number of hominid species. Will genetically modified human beings be ever considered as patentable subject matter? For the present, the evolving policies specifically bar the patenting of new genetic characteristics in humans. But as one official of the U.S. Patent Office remarked, the decision to patent higher life forms might eventually lead to the commercial protection of new human traits (Schneider, 1987, p. 15).

3. Disadvantages for the Developing Countries

The emerging biotechnologies are extremely powerful tools that could be used to help alleviate some of the most pressing problems in the field of health care, food security and energy supply in the least developed countries. However, control over these revolutionary tools is firmly in the hands of a few transnational corporations from the pharmaceutical, food processing and chemical sectors of the industrialised countries.

The emerging biorevolution, as it is being directed now, could further aggravate the problems of the poor by replacing their traditional export crops with laboratory-produced substitutes in the North. It could also aggravate evironmental destruction through further erosion of genetic diversity, as local, adopted crops are replaced by new high tech seeds, and increased use of harmful agrichemicals. Moreover, the privatisation of biotechnology by the North is effectively blocking Third World access to the basic resources, knowledge and techniques of the biorevolution. Ironically, much of these resources (genes), come from and will be further extracted from the developing countries themselves, as they harbour most of the planet's wealth of biological diversity.

II. A WORLD PATENT CONVENTION ON BIOTECHNOLOGICAL INVENTIONS

In the field of patent laws, there is a strong tradition of over a century of international co-operation. As international trade grew, the need for a harmonized protection of intellectual property rights was realized. Various international conventions were signed to regulate patent matters between member states. The beginning of this international co-operation on the protection of industrial property was the Paris Union Convention which is a universal treaty establishing certain basic rights for member countries. The original convention was signed by 11 countries. As of 1988, more than 90 nations were members of the Paris Union which is now administered by the World Intellectual Property Organization (WIPO) which was created by a convention in 1967. In 1974, the WIPO became the United Nations specialized agency. The European Patent Convention and the International Union for the Protection of New Varieties of Plants are another two landmarks in the direction towards the harmonization of patent protection (Beier, 1985).

These international and regional conventions on patent protection reflect mankind's gradual movement towards global co-operation and solidarity. These endeavours towards common policies and procedures reveal mankind's growing awareness of the world as a close-knit community whose members depend on each other for their well-being. Differences in the application of patent laws to a particular technology as between one country and another were the source of undesirable imbalances and conflicts. These steps towards greater harmonization were attempts to correct these situations. Mankind is now moving towards another important higher level of community-consciousness. We are entering a new era characterized by a sense of solidarity and co-operation with the community of the human species which includes both the present and future generations. This wider context of community-belonging points to the need of a reform in our legal system aiming to solve the conflicts and imbalances between generations.

Intellectual property protection of microorganisms, plants, animals and biological process is becoming of increasing concern to the world community. In view of the potential threats which the patenting of new life forms might have on the future evolution of all species, the world community is feeling the need of co-operation and co-ordination in the field of biotechnology. A number of differences still exist between nations regarding intellectual property protection for biological inventions. One such difference is the crucial issue of what constitutes patentable subject matter.

The need of a World Patent Convention on Biotechnological Inventions is becoming more and more urgent in view of the rapid advancement of genetic engineering. Governments of nearly all industrialized and many of the industrializing nations are formulating policies to support biotechnology and are increasing budgets to this end. In the near future, biotechnology will become a sizeable sector of the world industry.

A World Patent Convention on Biotechnological Inventions should be directed towards solutions that provide the best possible protection to future generations. International co-operation in the field of biotechnology is urgently needed for the well-being and interests of the present and future members of the human species. Since the patenting of new life forms can create conflicts and imbalances between present and future generations on the one hand, and between some parts of the world and others on the other, a World Patent Convention on Biotechnological Inventions should aim to foster solidarity among the present and future members of the human species.

III. EXPANDING THE COMMON HERITAGE

Patent decisions on microorganisms, plants and transgenic animals have been the target of harsh criticism by various scholars who all claim that a "suitable braking mechanism" should be invented in order to control the accelerating speed of biotechnological power (Kass, 1981, p.52). In what follows, I shall argue that the concept of the 'common heritage of mankind' seems to be that braking mechanism suitable enough to guide on the right track the emerging patent policies on biotechnological inventions. The application of the inspiring principles of this concept to a World Patent Convention guarantees to protect the well-being of the present generation of the developing countries as well as the long-term interests of mankind from the potential threats created by the granting of ownership rights on new life forms. The ethical and legal insights of this concept are a great source of inspiration to reconcile the human race and to put the law of solidarity and co-operation in place of the law of competition and self-interest.

1. Defining Genes as the Common Heritage of Mankind

The first international discussion about the common heritage of mankind principle initiated with the attempt to reform the traditional regulation of the Law of the Sea. Modern technology gave the advantage to a few

developed countries to explore and exploit the vast underwater and
seabed resources. In his statement of November 1, 1967, the Maltese
delegate to the United Nations, Ambassador A. Pardo, argued, in the
context of his proposal of a new Law of the Sea, that the concept of the
'common heritage of mankind' incorporates three main characteristics
(1975).

2. Non-appropriation

First of all, the concept of the common heritage of mankind is not a
theory of property. In fact, it implies the absence of property.[2] The common heritage engenders the right to use certain property, but not to own
it. Accordingly, the ownership of common goods are assigned neither to
mankind nor to any sovereign user. The key consideration is access to the
common resources, rather than ownership of it.

The aspect of non-appropriation is highlighted by the usage of the term
common heritage. The word common refers to a thing which belongs to
everyone, or which is shared in respect to title, use or enjoyment, without
apportionment or divisions into individual parts. The concept of heritage
conveys the idea that common things should be regarded as inheritance
transmitted down to heirs, such as estates, which by birthright are passed
dawn from ancestors to present and future generations. Since certain
goods constitute a heritage which is common to all mankind, it follows
that all present and future members of the human species, no matter
whether they are living now or in the future, have the right of access to
these common goods, without however claiming any right of ownership.

If there is an obvious component of the common heritage of mankind,
indeed more obvious than the resources of the seabed itself, it is the
genetic system. Nothing is so clearly heritable as the genes. The past,
present and future sharing of common genes is the most unitive factor of
all species. The genes of all existing species are inherited from their past
ancestors and will be transmitted to their future members. The genetic
system can definitely be considered as a true heritage because the genes
constituting a particular species are passed on from one generation to the
other. The genes present in every species are indeed common because all
past, present and future members share the same genetic linkage. Species,
in spite of their evolutionary process, remained a closed genetic system
precisely because their genetic heritage is uninterruptedly handed dawn
from one generation to the other.

This spatial and temporal interrelatedness within all species as a result of their common heritable genes applies also to the human species. Human genes are common to all past, present and future generations because the same genetic structure is inherited from one generation to the other. A human being is identified as such because of his membership of the human species. The collective human gene pool knows no national or temporal boundaries but is the biological heritage of the entire human species. Various developments in recent times have contributed to the growth in the awareness in mankind that it is a single, evolving subject, with the capacity to act and regulate itself collectively. Beyond doubt, more than anything else, progress in the science of genetics has contributed to the awareness of the physical continuity of mankind throughout time.

In order to understand the common genetic system of the human species and its physical continuity in time, it is important to clarify the concept of species. A species is distinct from a logical class (like the totality of "yellow" things, for instance). A species is a physical object which does not exist all at one time. It began to exist during some time in the past and will continue to exist in the future. A species occupies at any time a certain amount of space. To say that "X belongs to a species" is not to say "there are other like X", but "X is a fragment of a whole." In a species there is therefore a material continuity.

In the context of the principle of non-appropriation implied in the 'common heritage' concept, the understanding of the human species as a collectivity which encompasses both present and future generations is important for two main reasons:

First, if the meaning of term mankind denotes more than just the present generation and includes all generations yet to be born, then a wider dimension of the subject of human rights emerges. It is the awareness of sharing a common genetic structure that has enabled mankind to perceive itself most clearly as a collectivity of rights and responsibilities. The idea of mankind as the new subject of human rights needs to be understood within the context of the 'third generation of human rights' or 'the rights of solidarity' (Agius, 1988). Both the present and future generations, or mankind as a whole, have the right:

- to inherit a healthy and diversified genetic heritage. No generation or any other part of mankind has the right to appropriate any segment of the genetic heritage, no matter how small that segment might be;
- to preserve its own genetic unity; and

- to enjoy the benefits of biological diversity which is mankind's common heritage.

Secondly, the idea of collective rights of the human species to have access to the common heritage implies that every generation has obligations to mankind. As René Dupuy points out: "When we invoke the principle of the 'common heritage' as belonging to mankind as a whole in time as well as in space, we must understand by that, that present generations are accountable to future generations for this heritage" (1973, p.71). In other words, the concept of mankind as a unity implies intergenerational obligations. Responsibility means solidarity with whole community of the human species. Every generation has the obligation:

- to conserve the genetic heritage. Actions which threaten to diminish genetic diversity, impede the biological evolution of species, and impair any species' unity should be banned. Responsibility demands foresight in face of uncertainties;
- to enhance the common genetic heritage by employing those genetic engineering techniques which can eliminate deleterious genes, without endangering the unity and diversity of species; and
- to transmit the genetic heritage to the future while leaving asmany options open to unborn generations as possible. If between two alternative courses of action, one leaves a greater scope for future choices, it is preferable to the one which leaves less.

The current patent system which protects the right to private ownership of small segments of mankind's genetic heritage is far away from the ethical ideals of the common heritage of mankind principle. In fact, the current patent system seems to be still under the influence of the *res nullius* principle. Traditionally, there were two legal terms expressing the principles regulating the use and ownership of property: *res nullius* and *res communis*. *Res nullius* means that property belongs to no one. The legal inference is that such property becomes susceptible to appropriation or exploitation by anyone who is capable of carrying out those acts. The other legal concept, *res communis*, refers to property which is owned by no one and which therefore is rendered available for use by everyone. Lands or regions deemed to be *res communis* are thus not susceptible to exclusive appropriation by any private agent.

The common heritage of mankind has evolved and was introduced in international law precisely to check the anarchical and *laissez-faire* attitudes implied in the tradition of ownership and use of property. A critical situation, similar to the development of marine technology during the post-war era, has now emerged in genetic engineering. After the Second World War, scientific and technological revolution gave an immense power to some developed countries to explore and exploit the resources of the ocean. Nowadays, advances in biotechnology have opened up another region for exploration and exploitation. The earth's genetic system is now the new area for appropriation and private ownership. The cracking of the genetic code and the recent developments of genetic engineering techniques are enabling a privileged few to have an access to the immense natural resource of genes. This new region, still considered as *res nullius*, is becoming accessible only to those who are capable of carrying on biotechnological activities. The current patent system permits the ownership of genes to those who first explore the region and succeed to manipulate genes and create non-naturally occurring creatures. Once the inventor proves that his multicellular genetically engineered organism is not a product of nature, ownership rights are immediately granted. The patent law assumes that innovations proposed by inventors are, because innovative and useful for some, simply good for mankind as a whole. Due to this wrong perspective, the few who own biotechnology are selfishly exploiting genes to the detriment of future generations.

If a World Patent Convention on Biotechnological Inventions was inspired by the common heritage of mankind principle, then an international legal instrument would regard the earth's genetic system as an area which could not be owned, neither in whole nor in part, by any State, scientific community or a privately owned biotechnology company. Genes would not be considered as a subject of appropriation of any kind, neither public nor private. Sovereignty of genes would be absent.

3. Management on Behalf of Mankind as a Whole

The common heritage implies participatory management, not ownership of goods. Though goods which belong to the common heritage would be without any owner holding legal title in the traditional sense, an international administrative agency would assume responsibility for overseeing and regulating every activity conducted in the common area. Management includes also the supervision of the use of resources.

The common heritage of mankind principle demands the setting up of an international body which would have the power and the institutional means to regulate all activities in the sphere of genetic engineering. It

follows that instead of a multiplicity of actions dispersed between countries, often in pursuit of short term benefits and myopic to far-distant consequences, the access to the common genetic heritage would be organized in accordance to an international policy which takes seriously into account the interests of both present and future generations. Genetic resource management would obviously imply the management of biotechnology itself. Adequate (participatory) management of mankind's genetic heritage would include:
– conservation of the genetic heritage;
– protection of the species from genetic impairment;
– application of genetic engineering techniques which are beneficial to mankind; and
– transmission of a healthy and diversified genetic system.

Since every member of the human species has not only the right to inherit and enjoy a rich and diversified genetic system but also the right to share in the management of these resources, unborn generations must also be represented. Sharing in the international management of common genetic heritage on behalf of the interest of all mankind implies the participation of all members of the human species. Future generations would participate in the administration of the human genetic heritage and that of the ecosystem by being represented by a "Guardian" who will be responsible for reviewing all activities carried out in genetic engineering which may affect the welfare of posterity. The Guardian would be entitled to appear before institutions or biotechnological companies whose decisions could significantly affect the future of the species to argue the case on behalf of future generations, hence bringing out the long-term implications of proposed action and presenting alternatives. His role would not be to decide, but to promote enlightened decisions. The Guardian would thus face the burden of opposing the firmly established attitude of our civilization in discounting the future. He would therefore introduce a new dimension – time – in public policies traditionally confined to the present.

Management of the common genetic heritage would also include the surveillance of biotechnology so that it would be used exclusively for peaceful purposes. Biotechnology offers an amount of benefits for mankind. The same technology, however, can be used to construct weapons with a capacity for massive destruction. Unlike nuclear weapons, chemical and biological weapons are relatively cheap and simple to develop. They are, however, dangerous to handle and control. Their killing power is horrendous and they pose threats comparable to those of nuclear technologies. Indeed, they can destroy mankind's genetic heritage. A world patent convention on biotechnological inventions would prohibit the destructive use of genetic engineering for military purposes.

Management would also imply the global co-ordination and protection of gene banks. The speed with which the genetic erosion of agronomic species is taking place has prompted an international effort to conserve the reservoir of plant genetic resources of food crops and other important crops. Many countries have already set up gene banks in order the pre-serve the biological diversity of many endangered plant and animal species. Under the influence of the common heritage of mankind, an international regime would be responsible for the management of these gene banks.

4. Benefit Sharing by Mankind as a Whole

The common heritage implies sharing of benefits. The right of access to the common genetic heritage implies that any benefit derived from ge-netic engineering should be shared internationally. According to the common heritage principle, private agencies engaged in the commercial profit or private gain of biotechnology would be deemed inappropriate, unless they operate to enhance the common benefit of mankind. All the members of the human species would be designed the beneficiary of biotechnology, not simply all states or nations. Mankind does not refer to the interests, needs and aspirations of a particular segment of the human species; it includes all generations yet to be born.

Scientific research in the sphere of genetic engineering should be conducted for the benefit of the human species. Such research should be free and openly permissible, so long as the genetic heritage is in no way physically threatened or impaired. All research results would be made available as soon as possible to anyone who expressed genuine interest in them. According to the common heritage principle, genetic research would be conducted to benefit all present and future generations, not merely the state, or the government or private company which sponsored the research. Furthermore, the scientific fruits of genetic research would be freely and publicly exchanged in the hope of fostering greater scien-tific cooperation and more extensive knowledge about the genetic system.

The current patent system still reflects the philosophy of self-interest. Knowledge is exchanged under the condition that the inventor can control the commercial use of the knowledge – its profit potential. The contract formed by the patent law brings together a certain contradictory partnership and principles: self-interest and common good, the ownership of ideas and the shareability or publicity of thought. The patent law seeks to promote the common good by licensing private interest. An international legal instrument inspired by the common heritage ideas demands that the cumulative heritage of scientific knowledge should be used for the interest of mankind, not for personal gain. Mankind reaps greater good from immediate circulation of important ideas than he does by keeping them in secrecy for profit motives. Many scientists working in private and university research centres are actually doing this (Boonin, 1989, p.263). This attitude is impeding scientific evolution. The common heritage perspective on the use of scientific knowledge for the relief and comfort of all the members of the human species is similar to that ideal cherished by modern scientists. For instance, Descartes confessed that "as soon as I acquired some general notions in the area of physics ... I believed that I could not keep them concealed without greatly sinning against the law which obliges us to procure as best as we can the general good of all mankind." (1980, p.33)

It should therefore be recognized as an ethical obligation on the part of scientists to communicate knowledge publicly when it is relevant to decisions significantly affecting the future of species, to evaluate alternatives and the degree of probability that can be attributed to them.

In short, a World Patent Convention on Biotechnological Inventions inspired by the concept of the common heritage of mankind would incorporate the following ethical and legal principles:

- non-appropriation of genes;
- shared management of biotechnology on behalf of mankind as a whole;
- benefit sharing of scientific knowledge by mankind as a whole;
- exclusive use of biotechnology for peaceful purposes;
- conservation and protection of the genetic heritage for future generations; and
- some form of reward granted to those who invent or innovate something within the sphere of biotechnology.

CONCLUSION

The idea of a World Patent Convention on Biotechnological Inventions will not be realized unless mankind's whole attitude towards genetic heritage changes. A change in our attitude towards genes, however, implies a change in our attitude towards nature in general. We are still under the influence of the ideology of the industrial revolution which rests ultimately on the particular belief that man is the overlord of nature, and that nature is the servant of mankind. This spirit of ruthless domination over nature needs to be changed by a new *Weltanschauung* inspired by the philosophy of the common heritage of mankind. The notions of protection and conservation of all the earth's genetic heritage do not reside in a legal and institutional framework, however necessary they are. In the final analysis, these notions reside in the minds of men.

A sustained educational effort is therefore needed to change man's perception of nature and help him to take cognizance of his duty to adjust his mental and intellectual attitudes and behaviour towards the conservation and protection of the genetic heritage. A radical change in the current patent system presupposes a widespread awareness that we are an integral part of the genetic heritage. The way we look upon and treat genes is the way we look upon and treat ourselves. The conservation of the genetic heritage is self-conservation, and the impairment and damage of the genetic system is self-destruction.

The philosophy of the common heritage offers an adequate program of reform to change, first, men's mind and attitudes, then, the current patent law system which regulates biotechological "products." As the Anglo-American philosopher A.N. Whitehead claimed: "A philosophical outlook is the very foundation of thought and of society. The sort of ideas we attend to, and the sort of ideas we push into the negligible background, govern our hopes, our fears, our control of behaviour. As we think, we live." (1938, p.63). First we have to start thinking in accordance with the philosophy of the common heritage; then an international authority would be established with the power and the institutional means to conserve and protect the genetic system, recognized as belonging to the common heritage of mankind.

This would be a true bioethical achievement for the benefit of all mankind.

Faculty of Theology
Foundation for International Studies
University of Malta

NOTES

[1] During the Hearings, various witnesses, representing different national groups, expressed the public anxiety about the decision to patent life.
[2] The term "common heritage" has been variously interpreted to mean "common property" and "common sovereignty", confusion resulting in part by the U.S. Secretariat's translation of "common heritage" into French and Spanish as *patrimoine commun* and *patrimonio commun* respectively. The expression "common heritage" is much preferable to the term "common property." Property as we have it from the ancient Romans implies the *jus intendi et abutendi* (right to use and abuse). Property implies and gives excessive emphasis to just one respect: resource exploitation and benefits derived therefrom.

BIBLIOGRAPHY

Agius, E.: 1988, 'From Individual Rights to Collective Rights, to the Rights of Mankind', *Melita Theologica* **34**, 45–68.

Boonin, L.G.: 1989, "The University, Scientific Research, and the Ownership of Knowledge", In: V. Weil & J.W. Snappers (eds.), *Owning Scientific and Technical Information: Value and Ethical Issues*, Rutgers University Press, New Brunswick, NJ.

Beier, F.K. *et al.*: 1985, *Biotechnology and Patent Protection: An International Review*, Organisation for Economic Co-operation, Paris, France, pp. 26–29; 106–107.

Descartes, R.: 1980, *Discourse on Method*, trans. D.A. Cress, Part VI, Hackett Pub. Co., Cambridge, UK.

Dupuy, R.J.: 1973, *Collected Courses of the Hague Academy of International Law*, vol. I, A.W. Sijthoff, Leiden, The Netherlands, p. 71.

Kass, L.R.: 1981, 'Patenting Life', *Commentary* **72** (6), 45–57.

Pardo, A: 1975, 'First Statement to the First Committee of the General Assembly, November 1, 1967', *The Common Heritage: Selected Papers on Ocean and World Order: 1967–74*, University of Malta Press, Malta.

Schneider, K.: 1987 (April 17), 'New Animals will be Patented", *New York Times*.

United States, Congress House, Committee on the Judiciary. Subcommittee on Courts, Civil Liberties, and the Administration of Justice: 1988, Patents and the Constitution: Transgenic Animals: Hearing, 100th Cong., 1st Session, U.S. Government Printing Office, Washington, D.C.

Whitehead, A.N.: 1938, *Modes of Thought*, The Free Press, New York, NY.

ERIC T. JUENGST

SHOULD WE TREAT THE HUMAN GERM-LINE AS A GLOBAL HUMAN RESOURCE?

I. INTRODUCTION

The flashpoint for most current international discussions of policies to govern genetic engineering is the question of whether or not to allow scientists to pursue "germ-line gene therapy": genetic engineering interventions that would effect changes in people that would be passed along to their offspring as a natural part of their genetic inheritance. The question of whether to proceed with germ-line gene therapy is usually posed as a question about our obligations to respect the interests of future generations. In this paper I agree with this general interpretation of the issue, but dispute an assumption that many analysts make in attempting to take the argument further. In short, I dispute the assumption that the proper objects of our concern in contemplating human germ-line engineering are the descendants of the engineered.

Some argue that, in analyzing the problem of germ-line gene therapy from the perspective of future generations, the human germ-line should be considered as a special asset in the "common heritage of humankind," susceptible to the same forms of international stewardship as the planet's sea-bed or great works of art. Paramount in that stewardship is the goal of preserving the value of our species' genetic heritage for future generations, just as we would preserve our other universal goods. This, in turn, leads to calls to protect our common "genetic patrimony" against the plunder and abuse of those who would divert it to their own private ends, and to presumptive condemnations of the germ-line gene therapy proposals, as an infringement of the "integrity of the genome" and an abridgement of the rights of future persons to inherit a germ-line "free from any engineered alternations."[1]

I will call this line of reasoning the "Common Heritage (or "CH") View" in genomic science policy. The CH View has been spectacularly successful in public policy settings over the last decade, influencing positions advanced in Switzerland, Germany, Canada, The Council of Europe, UNESCO, France, Australia, and at any number of international conferences.[2] Moreover, it seems to flow naturally from the claims of

E. Agius and S. Busuttil (eds.), Germ-Line Intervention and our Responsibilities to Future Generations, 85–102.

conscientious molecular geneticists that, in some way, their work really is the public's business. Unfortunately, as an approach to science policy, the CH View is conceptually flawed and socially dangerous. I shall argue that, rather than (1) distorting accepted biological concepts and (2) risking social abuses in order to (3) enforce an awkward and impractical right to inherit one's share of a "common heritage of humankind," it (4) makes better sense and will be (5) ethically more imperative to pay attention to the rights and interests of those who will find themselves still coping with a natural ancestry in an increasingly engineered world.

II. WHY GERM-LINE GENE THERAPY SEEMS LIKE MINING THE SEA-BED

"If there is an obvious component of the common heritage of mankind, indeed, more obvious than the resources of the sea-bed itself, it is the human genetic system." (Agius, 1990, p. 140)

It is not hard to see why the human germ-line and the genes it carries are considered by so many to be special assets in the "common heritage of humankind." The human germ-line is the lineage of living cells that weaves itself through each of us, developmentally linking the zygotes at our origins to the gametes we produce after puberty. Since the zygotes at our origins are created by the fusion of other people's gametes, and since individual gametes of our own can, in exceedingly rare instances, fuse with cells from another person's germ-line to produce new zygotes, it is natural to think of the germ-line as a thread of living tissue that runs between parents and their children, linking each new generation to the last.

In fact, as long as one is willing to assume that our species grew from a single origin and has since been true to itself[3], the human germ-line might be interpreted as the golden thread of human biological life, the thread that connects all of us as one family, and through which we pass on that connection to our children, as a unique and universal human legacy. Moreover, since all the genes that underlie our individual human capacities have (traditionally) been transmitted to us through the fusion of our parents' germ-lines into our own, the value of the legacy of human germinal connection is not merely our common identity as members of the human family. As the transmitter of genes, it is the germ-line connection that allows us to inherit from our parents and endow upon our children the range of functions, capacities, and abilities that have enabled our species to become so successful in dominating Earth's ecosphere. As appealing as this story is, however, for the social values that it seems to embody, it needs to be abandoned as a basis for science policy.

III. WHY GERM-LINE GENE THERAPY IS NOT LIKE MINING THE SEA-BED

There are two sets of objections to evaluating germ-line engineering proposals through the lens of the CH View. The first are scientific and conceptual: framing the issues in terms of a "right to inherit a genetic pattern that has not been artificially changed" requires assumptions about human biology that make the effort incompatible with standard scientific thought. The second set of objections are ethical and political: grounding our moral concern for posterity in the CH view risks being self-defeating, by facilitating the very social abuses of genetic science that it seeks to prevent. The interests of posterity that are at stake in the genomic science policy decisions we make today are interests in social opportunities, not natural resources. Rather than fretting over biological risks of germ-line engineering we know nothing about, we should be fulfilling our obligations to the future by preventing the practice from exacerbating forms of genetic discrimination we already know all too well.

A. Scientific Conceptual Problems with the Common Heritage View

1. The Human Germ-Line
Is the human germ-line really a resource for humanity, like the sea-bed, the atmosphere, or even great works of art? It may be helpful to begin with some simple terminological clarifications. "The germ-line" is a concept from embryology and histology, and is used in those fields to label the lineage of dividing cells within an organism that link its zygote stage with its fully differentiated gametes. From the biologist's point of view, each organism's germ-line terminates in its gametes. The next zygote is not another cell within the parent organism's germ-line, because it is not the result of another episode of cell division within that developmental lineage. Instead, the zygote fuses the end-products of two germ-cell lineages, and founds a third.

This means that, strictly speaking, while individual humans have germ-line cells and germ-cell lineages, the human species does not have "a germ-line" in the geneological sense which the CH View requires. Moreover, it is important to biologists that the human species does not have a continuous germ-line, because that is precisely what distinguishes them as sexual eucaryotes from asexual organisms that persist as continuous lineages of mitotically dividing cells. In effect, human beings gave up their germ-line relationships long ago, in favor of sexual ones.

Of course, proponents of the CH View are free stipulatively to define the phrase "the human germ-line" for their own purposes, or to use the biological germ-line concept in an explicitly metaphorical way. The notion of a "human germ-line" as a transgenerational network with its own ontological integrity fits well – suspiciously well, in my opinion – with our pre-analytic cultural intuitions about human geneologies and family "blood-lines," and so is useful in setting the stage for the "common heritage" view. However, those who extend the concept this way should be aware that they are departing from the scientific understanding of the world when they do so. Biomedically speaking, there is no intergenerational "human germ-line" that could serve as an asset to the future, and efforts to ground social policy in such an idea may be greeted by the scientific community with the same scepticism they would give a policy that assumed that blood literally passes from parents to children in a familial "blood-line." [4]

2. The Human Genome

On the other hand, if our germ-lines do not literally link us to our families, the genes that are passed on during the merger of two germ-lines do. I have inherited (almost) exact copies of genes my parents carried, and will pass on (almost) exact copies of the same genes to my children. Perhaps it is that inheritance to which the CH View refers? Given that what the CH View seems to value about our "genetic patrimony" is its ability to give us our human capabilities, this seems plausible. It is the GENE-line, not the germ-line that is really at issue in germ-line gene therapy.[5] Perhaps it would be more correct, then, to say that it is the complement of human genes that each germ-cell lineage carries the "genome" that it shares with all other human germ-lines, that is at stake in the CH View. This is the way that UNESCO (1995) is now proposing to put it, in its draft declaration that: "the human genome is the common heritage of humanity." (Cf. UNESCO, 1995).

Does germ-line gene therapy endanger the rights of future generations to enjoy their rightful share of our common human genome, like undersea mining can deny them the fruits of the sea-bed? Again, the geographical metaphors that are used in discussions of "mapping the human genome" make it attractive – too attractive, in my view – to interpret the human genome like the sea-bed, as an explorable, exploitable, mutable "place." But we shouldn't be misled by our metaphors into reifying the concept of the human genome inappropriately. Unfortunately, if it is to continue to mean the same thing for science and public policy, "the human genome" is not something that can play the role of the sea-bed in evaluating germ-line gene therapy.

From the geneticist's point of view, the human genome is a heuristic abstraction, like the anatomist's concept of the "human skeleton." The "human genome" is defined as the full set of genetic loci that character-izes our species, together with the structural (noncoding) elements that connect them, just as the skeleton is defined as the full set of articulated human bones. In every individual, the gene occupying each genetic locus takes a particular molecular form, as one or another of the "alleles" or patterns the DNA can take in that locus: that is the source of the observ-able phenotypic diversity our genetic inheritance provides. The concept of the Human Genome, however, does not concern itself with all the substantive variations or different alleles that are possible at a given locus, any more than the concept of the Human Skeleton must account for the minute variations observable between the bones of different people. The human genome, unlike "the seabed," is a formal scientific concept, with no particular material referent in nature. That is why it does not matter to the Human Genome Project (unless we are quite mistaken about the nature of our species) whose chromosomes are studied by those conducting the genome mapping research.

Unfortunately, if the human genome is an abstraction and not a natural object, our worries about preserving its integrity for future generations become concerns about the future of an idea, not a natural resource. Thus we might legitimately assert the right of future generations to inherit the concept of the human genome, uncensored or misinterpreted, as part of their common scientific heritage.[6] But that would be to engage in a debate over the future of the stock of human ideas, not the range of human capacities. To that extent, it seems to be diverting us from our central concerns over the future effects of germ-line engineering.

Of course, we all have interests in preserving the integrity of our

personal specimens of the human skeleton, and it's plausible to argue that
we all have duties not to put the skeletons of members of future generations at risk (e.g., by using thalidomide recklessly). In the same way,
advocates for individual members of future generations could defend their
clients' right to inherit a complete complement of human genes, protected
from major loci deletions (or additions?). But they would not have to
appeal to the CH View to do so: it would be more effective simply to
argue that their clients had been directly harmed by the malpractice of the
genetic engineers, just as people already argue in "wrongful birth" suits
involving teratogens like thalidomide or accutane.

Moreover, interpreting the CH view this way, as defending every
individual's right to an intact genome, will not take us very far in evaluating most proposals for germ-line engineering. As long as the intervention
contemplated would only change alleles at a natural locus, rather than
deleting existing loci or adding new kinds of genes, new loci, the concern
to preserve the integrity of the human genome would be irrelevant to the
problem. For practical purposes, the genetic engineer can go a long way
towards both therapeutic and enhancement applications before it will
become necessary to contemplate inducing "knockout mutations" or
adding new loci in humans.

3. The Gene Pool

The fact that human genes do come in multiple forms suggests still another
refinement in the interpretation of the CH View. We could say, with Agius
(1990) and others, that it is humanity's stock of alleles – the sum total of all
our individual forms of each gene – that is really at issue in the CH view.
After all, it is here rather than in our common genomic plan or private
germ-lines, where the diversity that is often cited as the principal value of
our "genetic patrimony" is displayed. Within our species' complete collection of Mendelian alleles, polymorphic DNA variations, mutations and
"permutations," and functional configurations of genomic structure lie all
the biological capacities that allow us to flourish as humans in so many
different ways. Moreover, unlike the genome, this collection is real: it
consists of specimens of DNA spread across the germ-cells of 4 billion
organisms. Just like pieces of art in the world's museum's, these specimens
of DNA can, in principle at least, be collected, preserved, restored, destroyed, stolen, sold, shared, transferred, and combined: in short, they do
seem to be the sort of thing that, like the sea-bed, might be preserved and
cultivated for the benefit of future generations.

The scientific term in English for the collection of human alleles and genetic variants is the "gene pool." The concept of the "gene pool" comes to us from population genetics and evolutionary theory. It was introduced surprisingly late, in 1950, by Theodosius Dobzhansky, who used it to help establish a Mendelian definition of "species" as "a reproductive community of sexual and cross-fertilizing individuals which share in a common gene pool" (Adams, 1979). Dobzhansky seems to have coined the term by translating loosely from the Russian "genofond" or gene fund, a term used in the '20s by his mentors in Soviet population genetics. In fact, the gene pool is sometimes still referred to as "the aggregate genetic resources available to the population, its genetic reserves, on which it may draw in undergoing genetic change" (Adams, 1979). This fits very neatly – too neatly, it seems to me – with our intuitive cultural notions that humans have, thanks to investments made on our behalf through the "wisdom of evolution," accumulated a "genetic endowment" on which we might draw to meet new challenges, and over which we now have stewardship to manage a common inheritance.

Unfortunately, from a scientific perspective it is nonsense to extend the metaphor of financial management to the evolutionary process in that way. When evolutionary biologists use the investment metaphor to explain the dynamics of a gene pool, it is the unseen, stochastic hand of natural selection, not the species itself, which manages the species' gene fund. That is not simply because biologists usually ignore self-conscious organisms like themselves that can intentionally try to intervene in the process. It is because the evolutionary process is, by definition, an unmanaged and unmanageable one. This can be seen in several ways.

First, the great rhetorical advance Dobzhansky made by substituting "gene pool" for "gene fund" underscores the fluidity and flux of the evolutionary process. An endowment can be left untouched and inherited by new executors without necessarily increasing or decreasing in value. The frequencies of different alleles in a gene pool are constantly fluctuating, because everything that affects reproduction changes them, including the random genetic reconfigurations of the sexual reproductive process itself. As Dobzhansky says, "The gene pool is in constant motion; if a simile is desired, a stormy sea is more appropriate than a beanbag [or bank account]" (Dobzhansky, p. 201). Contemporary geneticists go even further, to suggest as metaphors phenomena that are more dynamic and have even less integrity than a pool, like rivers (Cf. Dawkins, 1995).

This dynamism means that it is no more possible to "manage" the human gene pool from the inside than it is for a pond to retain its integrity in the absence of a basin. Every human decision concerning reproduction "tampers" with the gene pool available to the next generation, and as a result there has been no "natural" gene pool to inherit since human beings started making reproductive decisions. From this perspective, humanity has committed an entire spectrum of "artificial interventions" in the gene pool, from the practices of celibacy and monogamy to contraception, adoption, and prenatal screening, to the care of people with genetic disease through their reproductive years. Unfortunately, left to its own resources, the CH view can give us no guidance for distinguishing within this spectrum between those practices which should count as illegitimate "tampering" and those that are consistent with future persons' rights to their fair share of the gene pool.[7]

Another way to put this point is to note that, despite the normative connotations of the financial metaphor, biology makes no value judgments about the kinds of evolutionary change that gene pools undergo as a result of natural selection. According to the architects of the CH View, one of the qualifications for being included within the "common heritage of humankind," is that the resources in question should be "such that they could be well managed only if they were managed on behalf of mankind as a whole" (Serracino Inglott, p. 195). However, from the evolutionary theorist's point of view, there are no "wise" or "poor" management decisions to make in attempting to manage the gene pool, because whatever changes occur will count equally as bona fide "evolution" in the species. Unlike the seabed or the atmosphere, there is no way to "deplete" or "improve" or even "preserve" the human species' "river of genes," because it has no settled natural state to use as a hemostatic benchmark.

Finally, recall that the gene pool is a concept that applies at the level of populations rather than individuals. Given their allegiance to traditional ways to transmitting genes to new generations, the proponents of the CH View cannot but argue that, in securing their right to "the transmission of the unique human genetic inheritance, free from any engineered alterations" individual members of future generations can demand access to the entire unabridged version of humanity's gene pool. At most, the CH View only supports the view that people should not be denied their rightful share of the gene pool's allele collection. But that introduces an evaluative distinction that quickly leads to harder questions. If it is not the species' entire collection of genotypes that individuals can claim as their heritage, how is their rightful subset to be determined? Surely not simply by reference to their own geneologies and the familial gene pools they can claim as their natural ancestry: that approach would quickly produce just the kind of genetic caste system that the CH view would seem to be designed to prevent!

The fact that the CH View cannot help its proponents to draw lines between acceptable and unacceptable genetic interventions and around the "rightful shares" of the gene pool that it seeks to guarantee, provides a hint that the answer to these problems lies in another direction altogether. In the end, I will suggest that it is not the genes one inherits that are important (or possible) to regulate in creating responsible genomic policies, but the social environment in which they will be expressed.

B. Ethical and Political Considerations

But perhaps I am too literal-minded about the scientific uses of the genetic concepts one finds decorating descriptions of the CH View. One could acknowledge the limits to the scientific senses of "germ-line," "genome" and "gene pool" in the debate over germ-line engineering, but argue that we should still listen to the metaphors that attend them. Even if evolution has no goals for humanity, humans can have goals for the evolutionary process, for both good and ill. So even if the effort is inconsistent with other policies and faces its share of practical and theoretical difficulties, might it not still be prudent to endorse the special use of the CH view in the case of human germ-line engineering, to provide a political basis for resisting the abuse of this potentially powerful technology?

Unfortunately, there is another set of objections that suggests that, even if we were to set aside the conceptual problems with the CH view, it carries serious social risks of its own.

1. Historical Warnings

First, it is worth noting that the CH View is not a new invention. As a paradigm for thinking about the social policy implications of genetics, it predates the Human Genome Project and our debates over germ-line gene therapy by 75 years. Mark Adams, a historian of genetics, demonstrates this nicely in his study of the history of our "gene pool" concept and its roots in the "gene fund" concept of the Russian geneticist, Aleksander Serebrovksy, in the 1920s. Adams points out that, as central as it has become as a concept of population genetics, the "gene pool" has its origins in an effort to help reconcile the science of genetics with the

Marxist ideology of the Soviet Union in the 1920s. Central to the effort
was a strikingly modern version of the CH View. Against those who
argued that Mendelian genetics suffered an inherent Social Darwinist bias
in favor of the capitalist elite, Serebrovsky argued that:

"If we consider our population, the citizens of our Union, we can regard them from one point of view as a group of subjects with full rights
who exercise their right to create their own happiness on earth, and
from another point of view, we can look at their totality as our social
treasure, just exactly as we look upon the total amount of wheat, milk
cows, and horses which create the economic power of our country. Our
country prospers not only because wheat grows and cows give milk,
but also because it has people who produce work of a certain level of
quality. This question is especially important when we move to the
"higher" levels of human creativity, to artistic, scholarly, and scientific
activity, to administrative work and a whole series of other manifesta-
tions of human nature. And if these elements actually rest on a basis of
heredity, then we have every right to look upon the totality of such
genes which create in human society talented outstanding individuals,
or to the contrary idiots, as national wealth, a gene fund, from which
society draws its people. It is clear that, not only can we not close our
eyes to our gene fund, but to the contrary, we must see if there are
processes operating within the gene fund which are changing it, and if
there are, to what extent it is for the better or worse...In order for the
reserves of various genes in a given locality to be properly managed,
we must look upon this stock as a kind of natural resource, similar to
reserves of oil, gold, or coal, for example" (quoted in Adams, 1979).

Serebrovsky then went on to advocate a series of eugenic proposals to
preserve and improve the glorious Soviet gene pool. The centerpiece to
his scheme of "Soviet Eugenics" was a plan to regulate human reproduction for the benefit of future generations. He argued that:

Children are necessary to support and develop society; children must
be healthy, able and active, and society has the right to ask questions
about the quality of the output in this area of production. We propose
that the solution to the question of the organization of selection in humans will be the widespread induction of conception by means of artificial insemination using recommended sperm, and not at all necessarily from a "beloved spouse" (Adams, 1979).

Unfortunately, in the context of Germany' s reviving power and militant eugenic policies, these proposals did not help Serebrovsky's career in Stalinist Russia, and Adams reports that the term "geno-fund" disappeared for almost two decades before being revived, shorn of its explicitly eugenic associations as "the gene pool."

Meanwhile, of course, much the same kind of language was used by eugenicists all over the world to help exploit the power of the CH View to advance political policies of social exclusion. Although these eugenic efforts did not have the benefit of Serebrovsky's "the gene fund" (they used "the germ-plasm" instead), they were often much more successful. In the U.S., immigration restrictions against people from the Mediterranean, prohibitions of inter-racial marriage, and the involuntary sterilization of 60,000 "feeble-minded" people were all justified in terms of protecting the integrity of the genetic stock on behalf of future generations. In some circles today, one can still find people decrying the long-term "dysgenic" effects of modern medicine in just the same way.

In rehearsing this history, I do not intend to imply that proponents of the CH View in the context of the germ-line debate have an underlying eugenic agenda, nor that the "gene pool" concept as it is used today is tainted in some way by its historical origins. Clearly, most proponents of the CH View endorse it precisely in order to forestall the possibility of new eugenic programs. However, I do think that Serebrovsky's invocation of the CH View to advance his eugenic agenda does display the ways in which rectifying the "gene pool" as a natural resource can open the door to unacceptable social policies just as easily as over-eager attempts to privatize and improve upon our nature.

Serebrovsky's invocation displays three ways in which the CH View can prepare the ground for unjust social policies and attitudes: it can encourage genetic reductionism, coercive reproductive policies, and discriminatory social attitudes.

2. Genetic Reductionism

By putting the human gene pool in the same normative category as the human stock of ideas, the CH View places an inordinate amount of weight on the role of genes in human flourishing. It centers our expectations about each others' accomplishments and our suspicions about each others' faults on our genes. To do so is uncomfortably reminiscent of the role we used to give the soul in human affairs, and the forces that were thought to affect it. But reducing the character and accomplishments of an individual to their genotype is a very narrow perspective on human nature, that most biologists would reject: genes are very far away from most of the action that makes human beings interesting to live with. When this kind of reductionism is combined with an ethic of personal responsibility for health, it can also serve to saddle individuals in future generations with private genetic responsibilities for problems which could have been addressed today as public (e.g., environmental) issues, by relocating the cause of the problem in their genes (Duster, 1990).

3. Coercion

Moreover, unlike protecting the sea-bed or atmosphere, no interventions can be performed on the gene pool without interfering with the lives of individual human persons. If the global managers of the genome were to decide that the dynamics of the human gene pool required regulation, how would they go about it? Just as Serebrovsky's eugenic proposals foundered on their invasion of the marriage bed, any form of global "management" of the gene pool would require invasions into the sphere of reproductive privacy. When this view is combined with a "genetic naturalism" (Knoppers, 1991) that eschews interventions into the natural gene pool, one is left only with reproductive interventions, inappropriately transforming what should be strictly personal issues, into problems for public policy (Duster, 1990).

4. Discrimination

Serebrovsky' s egalitarian opponents were offended by the tendency of Western geneticists to base their eugenic theories on elitist views of human classes and races. Note, however, that even in sketching a "Soviet Eugenics," Serebrovsky makes normative distinctions between different human types, with those involved in "scientific and administrative activities" at the top of the list. Genetics is the science of human differences, and as long as proponents of the CH View admit that some items in the human species allele collection are more useful, pleasant or desirable than others, the door will be open to classify the members of future generations according to the value of the share of the human patrimony they will have been allotted.

III. ANOTHER PERSPECTIVE:
INSURING EQUALITY OF OPPORTUNITY WITHIN FUTURE
GENERATIONS

If the notion of a common human genetic heritage is so murky and un-helpful, why is it that germ-line gene therapy strikes us so forcefully as a technology which bears on the interests of future generations, and for which global management would be appropriate? To see the real issue that this technology raises, it is helpful to think about what the purposes of a germ-line genetic intervention could possibly be.

A. Phenotypic Prevention

The first possible purpose for a germ-line intervention would be a tradi-tional medical purpose: to prevent the manifestation of a disease by treating it in an affected individual's pre-embryo phase or in the gametes of the patient's parents. This might be called "phenotypic prevention," since it seeks to prevent the expression of a disease phenotype. Here, germ-line gene therapy enjoys the support of the medical model, and even the most conservative official statements have started to waffle toward making exceptions to allow it, even if they have to import additional moral principles to do so. In this situation, any effects of the intervention on subsequent generations can be rationalized as unintended and indirect side-effects of the therapeutic intervention, and placed in the same ac-ceptable category as the many other ways in which people influence (or "tamper with") their germ-line cells in the course of normal life (Lappe, 1991). Of course, if these germ-line interventions start achieving their preventive goals by enhancing some health maintenance system within the patient's body – just as somatic cell gene therapies for hypercholes-terolemia already are doing[8] – this class of interventions may begin to seem more problematic. But given our present focus on issues relevant to future generations allow me to move on.

B. Genotypic Prevention

A second, and more controversial, possible purpose for germ-line gene therapy would be what one might call "genotypic prevention": the effort to use the practice to try to reduce the incidence of a particular genotype in the next generation of a population. This is the application in which the interests of future generations are clearly at stake. In order to achieve this goal, germ-line interventions would have to be carried out at a mass scale, as part of public health campaigns involving as much of the population as possible. This would involve a public commitment to the value judgment that the targeted genotypes are disvaluable enough to warrant efforts to exclude them from the population entirely. Inevitably, the success of such programs would be evaluated in terms of their abilities to reduce the frequency of the targeted alleles in the population, and the resulting savings in cost and morbidity.

Clearly, one set of issues that any such use of germ-line therapy would face is the issue of balancing the public health savings of such a program against the liberty of those individuals asked to participate. Prospective parents are likely to resist efforts to police their germ-lines, and health professionals are likely to resist efforts to tie funds for the services to their success in having their patients make decisions the State' s way. But again, these are issues for the first generation faced with this prospect. What about the future?

As advocates for people with disabilities continue to point out, there is another constituency that would be put at risk by the public health use of germ-line engineering: the community of people who will continue to carry the offensive genes during the life of the public health program – the unengineered. The further into the future this use of genetics persists, the more their interests are damaged, in two different ways:

A. Social "Handicaps"

First, if it is true that those who are freed from the "pathogenes" they would have inherited by the public health campaign will actually enjoy a greater range of capacities than those who continue to carry their "genetic load," then those whose CH rights have been respected (by oversight on the part of the public health authorities or zealous conscientious objection to the program) will be at a distinct social disadvantage. From the per-spective of their genetic heritage of course, they will continue to be in fine shape, having inherited just their rightful and natural share of the gene pool. However, their relative capacities, and thus their relative share of the range of human opportunities available to their population will have been decreased. They would be in the same position as people from the 4th world who lack antibiotics are today. The fact that their people have not traditionally had access to antibiotics, does not mitigate their claims to gain their fair share of that social good and the opportunities it can provide for human flourishing. It is in this way that germ-line gene therapy risks the creation of a real genetic "underclass" where there was none before.

B. Social Stigmatization

Moreover, because of the genetic reductionism that the CH View encourages, those who are left behind in a public health program designed to defend the population against some genetic disease are at risk of having whatever real disadvantages they may face greatly exaggerated by overly deterministic readings of their genotypic "handicap." Because a single genotype can almost never predict with certainty the burden of suffering someone will experience, and can certainly never sum up the overall worth of the person, any social discrimination on the basis of "retrograde" genotypes will be unjustified in the vast majority of cases. Advocates for pathogene carriers can point out that, by focusing public health measures on genotype as a cause of human suffering, the germ-line engineers risk stigmatizing gene carriers as vectors of suffering and "irresponsible reproducers" of the genetic underclass. In this context, germ-line gene therapy seems less like mining the seabed, and more like introducing a culturally biased intelligence test into our school systems: it could function to exaggerate the advantages that some portions of the population already possesses, and thereby widen the gap between the opportunities available to them and to the rest.

If I am correct that the most urgent long-term social policy issues to flow from the possible uses of germ-line gene therapy actually involve the interests of the un-engineered, then it should be clear that something more than the CH View will be required for the creation of responsible public policy in this area.

Notice that even in the beginning, no one has assumed that the entire "gene fund" was valuable to our species: Serebrovsky acknowledges that there are genes for idiots as well as geniuses in the Soviet national treasure. Some alleles do seem preferable to others at many loci. By what standard are these intuitive judgments made? Without offering the full argument for it here, let me just propose what the subtext of this essay has been suggesting all along: that to the extent that it is the range of human capacities that is what is valuable in our common genetic heritage, the critical underlying principle that needs to be brought to the surface in contemplating germ-line gene therapy is the role of opportunity in human flourishing. We seek to restore normal capacities (and perhaps build new ones) to give ourselves more opportunities for the experiences that give our lives meaning: savoring relationships, creating art, meditating on nature, etc. In other words, in prizing our genetic heritage and evaluating how to cultivate it, it is the opportunities for creativity that it makes possible, not the genes themselves, that we want to preserve. To do so means paying more attention to the social context of future generations than to their genetic resources.

IV. CONCLUSION:
ADDRESSING THE SOCIAL OPPORTUNITY INTERESTS OF FUTURE GENERATIONS

In conclusion, allow me to summarize what I think are the most important social policy goals in attempting to prepare for the future of genetic technologies (with the interests of future generations in mind).

First, our emphasis in designing public policy about the genome should not be on preserving some typological 19th-century notion of the "image of man." Instead, the emphasis should be on providing and protecting a range of opportunities for men and women to flourish. In the long run, that means that those who benefit from gene therapy – or even genetic enhancement – are not the ones we should be concerned about. Rather, the preeminent need will be to protect those among the future generations "unfortunate" enough to enjoy an untampered genetic inheritance from the social discrimination and the unfair disadvantages that they could face in living and working with their genetically engineering neighbours.

The outline of those policies are already clear in the policies that are already emerging to secure the opportunities due people with disabilities. They take both negative and positive forms:

1. We should eschew proposals to use germ-line gene therapy – or, for that matter, genetic screening techniques – as public health tools to reduce the incidence of genotypes associated with genetic disease and disability in the population, as if particular genotypes represented a form of expensive "pollution" that could and should be cleansed from the gene pool. This way of thinking reduces the identities of people with disabilities to their genotypes, and encourages others to discriminate against them.

2. We should continue to expand the social opportunities and protections available to people with genetic differences and disabilities, by working to prohibit unwarranted discrimination on the basis of genotype alone.

3. Finally, we should strive in our rhetoric about the interests of future generations to focus on the promises we would like to make to our children, rather than fret about what we have (or have not) inherited from our parents. The human gene pool, unlike the sea, has no top, bottom, or shores: it cannot be "used up." The reservoir of human mutual respect, good will and tolerance for difference, however, seems perennially in danger of running dry. That is the truly fragile heritage that we should seek to preserve in monitoring genetic research on behalf of the future.

School of Medicine,
Case Western Reserve University
Cleveland, Ohio, U.S.A.

NOTES

[1] One of the clearer statements of this view is to be found in Agius (1990):

"The genetical relation among all generations is one of the most unitive factors among the community of the human species. Human genes are common to all generations. The collective human gene pool knows no national or temporal boundary, but is biological heritage of the entire human species. They are a common heritage because they are handed down from one generation to another. In a relational context, the rules of inheritance assume a much broader perspective than those ususally recognized on the individual level. In fact, ownership rights are from the viewpoint of mankind as a whole... no generation has therefore an exclusive right of using germ-line therapy to alter the genetic constitution of the human species" (Agius, 1990, p. 140).

[2] See Mauron and Thevoz (1991) and de Wachter (1993) for reviews of these statements, as well as UNESCO (1995) and Foundation for International Studies (1994).

[3] E.g., no cross-species contributions from neanderthals or the nephileme.

[4] There are, of course, millions of particular lineages of germ-line cells in individual human beings. Strictly speaking, then, to express concern about the effects of pre-embryo transformation or gametocyte therapy "on the germ-line" means worrying about the risks of the intervention to the health of the individual transformed: i.e., the risks of iatrogenic mutagenesis, or over expression, or epigenetic complications that are already familiar to those working with somatic cell gene therapy. These are real concerns, but not concerns that flow from the communal nature of the germ-line, or risks that are likely to have a widespread impact on future generations requiring international management. Existing science policy governing biomedical research practices already insures that germ-line gene therapy will never emerge

as a legitimate clinical tool unless those basic concerns can be satisfied first.

[5] As long as one is willing to disregard any possible cytoplasmic modes of inheritance!

[6] Perhaps we could even defend that right against those who would pass on the garbled notion of the human genome as a natural resource!

[7] Of course, most advocates of the CH View do go on to make such distinctions; but they have to do so by importing other moral principles into their analysis. Thus, as an exception to its ban on "tampering" with the genome, the Council of Europe allows all practices which can be otherwise justified "in accordance with certain principles which are recognized with being fully compatible with respect for human rights" (Council of Europe, 1982). Similarly, Agius appeals to principles derived from his Whiteheadian ethical theory to explain how he can support some uses of germ-line genetic engineering, despite his commitment to the CH View (Agius, 1990). Of course, in importing these principles to settle all the practical questions about which interventions are acceptable, they render the CH View and its prima facie injunctions superfluous!

[8] By increasing the patient's number of low density lipo-protein receptors beyond their normal level. See Wilson, 1992.

BIBLIOGRAPHY

Adams, M.: 1979, 'From "Gene Fund" to "Gene Pool": On the Evolution of Evolutionary Language', *Studies in History of Biology* **3**, 241 – 285.

Agius, E.: 1990, 'Germ-line Cells – Our Responsibilities for Future Generations', in S. Busuttil, E. Agius, P. Serracino Inglott, and T. Macelli, (eds.), *Our Responsibilities Towards Future Generations*, Foundation for International Studies, Valletta, Malta, pp. 133 – 143.

Council of Europe (Parliamentary Assembly): 1982, *Recommendation 934 (1982) on Genetic Engineering*, Council of Europe, Strasbourg, France.

Dawkins, R.: 1995, *River Out of Eden: A Darwinian View of Life*, Basic Books, New York, NY.

de Wachter, M.: 1993, 'Ethical Aspects of Germ-line Gene Therapy', *Bioethics* **7**, 166 – 178.

Dobzhansky, T.: 1970, *Genetics of the Evolutionary Process*, Columbia University Press, New York, NY.

Foundation for International Studies: 1994, 'World Declaration on Our Responsibilities Towards Future Generations', in E. Agius and S. Busuttil, (eds.), *What Future for Future Generations?*, Foundation for International Studies, Valletta, Malta, pp. 305–311.

Knoppers, B.: 1991, *Human Dignity and Genetic Heritage,* Law Reform Commission of Canada, Montréal, Québec.

Lappé, M.: 1991, 'Ethical Issues in Manipulating the Human Germ-Line', *Journal of Medicine and Philosophy* **16**, 621 – 641.

Mauron, A. and Thevoz, J.-M.: 1991, 'Germ-line Engineering: A Few European Voices,' *Journal of Medicine and Philosophy* **16**, 649 – 666.

Serracino Inglott, P.: 1994, 'The Common Heritage of Mankind and Future Generations', in E. Agius and S. Busuttil (eds.), *What Future for Future Generations?* Foundation for International Studies,Valletta, Malta, pp. 193–199.

UNESCO (International Bioethics Committee): 1995, 'Revised Outline of a (UNESCO) Declaration on the Protection of the Human Genome,' *Eubios Journal of Asian and International Bioethics* **5**;4, pp. 97–99.

Wilson, J., *et al.*: 1992, 'Ex Vivo Gene Therapy for Familial Hypercholesterolemia', *Human Gene Therapy* **3**, 179 – 222.

PART IV

SOCIAL RESPONSIBILITIES OF GENETICISTS
TOWARD FUTURE GENERATIONS

QUI RENZONG

GERM-LINE ENGINEERING AS THE EUGENICS OF THE FUTURE

This paper consists of three parts. Part I challenges the "rights approach" to the future generations problem. Agreeing with D. Parfit's argument that an appeal to rights cannot entirely solve the future generations problem (called the Non-Identity Problem by him), further possible arguments will be explored: the talk on rights of unconceived people is difficult, which rights or whether rights would be claimed by future generations is unknown, and the talk on rights of future generations is only a weak voice in the conflicting voices of a huge chorus of right-talks.

In Part II, it will be argued that it is necessary to have a "paradigm-shift" or establish a new conceptual framework to deal with the future generations problem for the individualistic framework prevailing in the West, on which the right approach is based. It is time to reject the self-interest theory which is deemed as rational in the West but as self-defeating and often not moral as Parfit argued. In the new framework, individual or self is not and cannot be independent from others, both in space and time. Self should be replaced by self-in-relations or relational self. This is why Confucius claimed that we should love and care for others. This is the difference between a human being and an animal; actually, even an animal does not always care for itself only. From a Confucian standpoint it is our responsibility to leave good and not harm to future generations, even if that will reduce the happiness or interests of present individuals.

In Part III, it will be claimed that there is a moral borderline between treatment and enhancement which applies to future generations. "Playing God" or positive eugenics should always be rejected. Underlying the enthusiasm to apply gene engineering for the purpose of enhancement is a misleading theory of biological or genetic determinism, and reflects a tendency to medicalize social problems. It also imposes a value system upon future generations, such that unpredictable risks or harms to future generations will far exceed our expectations.

E. Agius and S. Busuttil (eds.), Germ-Line Intervention and our Responsibilities to Future Generations, 105–116.

I. THE RIGHT APPROACH TO THE FUTURE GENERATIONS
PROBLEM

When I read the publications on the future generations problem, I was deeply impressed by the overwhelming use of rights talk, the same as in other fields. Definitely, rights talk of future generations has been and will be a help for us to concern ourselves with future generations (Busuttil, 1990; Agius, 1990a; Agius, 1990b; Agius, 1990c; Agius, 1994; Serracino Inglott, 1990a; Inglott, 1990b), but some doubts have been cast on it. For example, Derek Parfit (Parfit, 1984, Chapter 16, p.124) in Reasons and Persons raised the question: "Can we wholly solve our problem by appealing to people's rights?" His answer is negative. Parfit's strategy puts forward first a Non-Identity Problem: how to respond to a 14–year-old girl's choice to have a child. Strictly speaking, this is a next generation's problem, not the future generations problem. However, Parfit applies his argument also to the remote future generations problem, as discussed in his *Repugnant Conclusion, Absurd Conclusion and Mere Addition Paradox* (Parfit, 1984, Chapters 17–19). The claim which the supporters of the positive answer may advance is: "The objection to this girl's decision (to have a child now not later) is that she violates her child's right to a good start in life." The possible counter-arguments Parfit provided are:

(CA) Even if this child has this right, it could not have been fulfilled.

This argument is plausible, I think. The child who is borne by this mother will be in either of two outcomes:

(c1) If it is borne by the mother when she is 14 years old, it will be this child and it will not have a good start in life.

(c2) If it is borne by the mother at her mature age, it will be that child, not this child. In this case, there is no start at all to this child.

As Parfit argues, this child could have said: "I waive this right" (Parfit, 1984, p.364]. Because if it waives the right, it could have a start which, though not good, is worth living. However, if it keeps the right, it could have no start at all, and consequently no right at all for it.

Parfit's counter-argument is important and plausible, but not adequate. I will add some other counter-arguments as to why an appeal to rights cannot wholly solve the future generations problem.

(ACA1) The talk on an unconceived child's right is difficult.

If an unconceived child has the right to a good start in life, it should have a more important right to have a start. So the rights talk is self-defeating. When pro-life activists argue against the use of contraception, they say that it violates the right of an unconceived child to life. For if no contraception is used the ovum will meet the sperm and be fertilized, and then conceived and delivered. Obviously, "unconceived child" is a self-contradictory word. If it is "unconceived," there could be no child. If there is a child, it has been conceived. Actually,the talk is on the right of an ovum or sperm rather than a child's right. It is absurd to say hundreds of ova or billions of sperms have rights.

(ACA2) Which rights or whether rights would be claimed by future generations is unknown.

Rights-talk is a recent event compared to human history. In non-Western countries there has never been rights-talk before they met Westerners. Even now, in many third-world countries rights-talk is not so stressed as in western countries, or has a different focus. Nobody knows the views to be held by future people. Possibly, they might hold a different view on rights from ours, or they put different stress on the priority of various rights, or even they reject any rights-talk because they may accept Confucianism. Now we say that future generations should have such and such rights, but many centuries into the future people might say we don't want such and such rights, or even say, we are not interested in rights. As Parfit argued, they could say: "I waive this right" or even say "I don't want this right." In other words, the value or belief system that future generations will hold might be quite or even widely different from ours.

(ACA3) The talk on rights of future generations is only one weak voice in the conflicting chorus of rights-talk.

Now there are many rights that we talk about. These rights are competing and conflicting. Many rights are being talked again and again, but never fulfilled. One reason is that these rights are in conflict. If this right is fulfilled, that right will not be. The second reason is that there are so many talks on rights, but very few talks on those who have the obligation to provide products or services to fulfill these rights. In the United States, people are talking about the right of a defective newborn to life. However, to support the life of a newborn with a very low birth weight, say 500 g, costs half a million U.S. dollars. But there is little talk about those who have the obligation to pay. In a context of so much talk on mutually competing and conflicting rights, the voice on those already born is always louder than the voice of those in favor, of the unborn, and unconceived. I do not mean that that should be the case, but that's more likely to happen.

An objection to this argument might be the election of a guardian as a representative and spokesperson of future generations; he or she could strengthen the voice on the rights of future generations. This practice might be helpful to future generations. But the problem persists. We don't know whether the one we elect as a guardian of future generations will be accepted by them or not, and how legitimate is this procedure. Even if a guardian were to be elected in the United Nations, her or his work would be very difficult. The status of a guardian for future generations is easily challenged by somebody and his/her work will be frustrated. For example, a guardian for future generations might challenge the Chinese policy of developing the auto industry, on the basis that this policy will create an environment in which future generations cannot live. His/her voice will be dampened by a billion voices who will benefit from the policy, including Chinese, Japanese, Western business persons, and a huge number of Chinese laypersons who are eager to have a car like the Japanese and Westerners. They will argue that this policy will benefit future generations, and that environmental problems can one day be prevented or solved by advanced and sophisticated technology. And representatives from the third world countries might suspect that the guardian is a real spokesperson of developed countries who wants to block the social development of developing countries.

> (ACA4) Rights-talk is based on individualism and related with self-interest theory which is self-defeating.

Although Parfit argues that the self-interest theory is self-defeating (Parfit, 1984, Chapters 1, 4, 9), he uses it to raise the future generations problem. His Non-Identity Problem means judging the choice from the perspective of this child or that child's interests. The same for his *Repugnant Conclusion, Absurd Conclusion and Mere Addition Paradox.* However, the future generations problem cannot be reduced to this or that child's problem. So we can say, that the presumption of Parfit's argument is still in the conceptual framework of individualism. However, the future generations problem is a problem of humankind. Humankind is not merely a sum of individuals. Any solution based on individualism won't prevail and will be self-defeating. One valuable approach would be to shift the focus to the rights of humankind rather than to the individual or to collective rights (Agius, 1990a). In this approach the conflicts between rights of humankind and individual or collective rights have to be appropriately resolved. And when we talk of the rights of humankind we will have to put more emphasis on duties or responsibilities.

II. THE CONCEPTUAL FRAMEWORK APPROPRIATE FOR RESPONSIBILITY TOWARD FUTURE GENERATIONS

There are many excellent writings on responsibility towards future generations (Busuttil, 1990; Agius, 1994; Bell, 1995; Gunatilleke, 1990; Serracino Inglott, 1990a; Serracino Inglott, 1990b; Macelli, 1990; Streeten, 1990; van der Veken, 1994; Weiss, 1990). I claim that the key to the future generations problem is the sense of responsibility of the present generation toward future generations. Now the problem is: whence comes this sense of responsibility or the responsibility itself? It may be argued that the responsibility for future generations comes from their rights. The difficulty of this argument has been shown in Part I. It can be further argued, that this responsibility comes from the disadvantages of future generations (Macelli, 1990). It is plausible. But for the individual who is pursuing his self-interest in a way that is deemed rational from the western point of view, do the disadvantages of future generations bother him? I will argue that the change from depletion policy to conservation policy needs a paradigm-shift or a change to a new conceptual framework to deal with the future generations problem.

From the historical point of view, the concept of right is based on the conceptual framework of individualism and contractualism. When people are subordinated to, or exploited and abused by somebody, to treat them as independent, autonomous individuals will justify their liberation or freedom from their exploiter or oppressor. This is what Locke, Hobbes, Rousseau, Kant and others have seen. It is still helpful to people who are exploited, oppressed, disadvantaged, and vulnerable nowadays. However, the individualistic framework on which the right approach prevailing in the West is based is completely inadequate to solve the future generations problem. For individualism pursuing self-interest is rational, but the action for pursuing self- interest may either benefit or harm others, or it may even be neutral. Parfit argued that the self-interest theory is self-defeating and in conflict with morality (Parfit, 1984, Chapters 6,14).

For the individual who is pursuing his own interest, others seem to have only an instrumental value to him. How can he care for others? Let alone the remote, unborn, or even unconceived future generations. When the action of the present generation affects future generations, individualism is even more inadequate.

If based on individualism the society is contractual. Each member of this society is a stranger to others, and is rationally pursuing his own self interest. The only mechanism which can protect each member's self interest without harm from others is a consenting contract between them. In the contract each partner has his/her rights and corresponding obligations to the other side. There is no obligation outside the contract. It follows that the present generation has no obligation to future generations because all these unborn or unconceived people cannot be a partner with present people in a contract.

It is time to reject the self-interest theory together with the individualism which implies it and the resulting contractualism, and instead adopt a new conceptual framework. In the new framework individual or self is not, and cannot be independent from others, both in space and time. An individual only exists and develops in relation with others. Self should be replaced by self-in-relations or relational self (Agius, 1990b; Held, 1993, pp. 57–63). It is why Confucius (Chan, 1964) claimed that we should love and care for others. For Confucianists, care for others is called *ren*, humanness. Where do you come from? From your parents. So the filial piety is the beginning of humanness. Filial piety is a human being's passion and first responsibility towards others. From this start, you should extend your responsibility and passion to your sisters and brothers (fraternity), your children (kindness), your spouse (fidelity), your friends (sincerity), your patients (compassion), your countrymen (harmoniousness), and foreigners (peacefulness). Because the birth and development of an individual owes so much to others, he has the duty to care for them.

Care for others is an indispensable requirement of a human being. As Confucianists argued, what differentiates a human being from an animal is that the former cares for others, but the latter cares only for itself. Actually, even animals do not always care for themselves only. Although we were born as human beings biologically, it is when we learn to care for others that we become human beings morally. As for future generations: "present generation plants trees, future generations enjoy the cool." That is our duty. So the care for, concern with, and responsibility towards future generations as well as the remote disadvantaged or vulnerable is a natural extension of ren, humanness. It is our responsibility to leave good and not harm to future generations, even if this will reduce the happiness or interests of present individuals. The sense of responsibility towards future generations hardly comes from future individuals' rights, but from the consciousness of humankind as a whole. An individual may exist or disappear, but humankind still exists. An action affecting negatively an individual or an action that proves positive for the individual, might be negative for humankind. It is wrong to sacrifice the present for an illusory happiness of future generations, but it is morally obligatory to reduce present luxurious consumption in order to avoid real harm to future generations.

In the non-individualistic framework the relation between human beings is not dominated by contractualism. Contractual relations may be appropriate in certain limited contexts, such as in business. However, the dominating relation between human beings is not contractual (Held, 1993, pp. 192–214). We cannot say that the relationship between parents and children, sisters and brothers, friends, co-workers, neighbours, physicians and patients, teachers and students, lawyers and clients, ministers and believers are all just a relationship of strangers in a contract where each one only has an instrumental value. Future generations are not one side in a contract, but part of mankind, just as past and present generations are.

III. GERM-LINE ENGINEERING AS THE EUGENICS OF THE FUTURE

Germ-line engineering is a sample which illustrates how our present action will affect future generations. Before we discuss the ethical issues in human germ-line engineering, I think we first should address the issues on how and why we use genetic knowledge, i.e., the values and presumptions underlying the use of genetic knowledge.

There will be no dissent on the answer to the "why" problem: we use genetic knowledge to improve human existence and quality of life. The following values can be promoted by the use of genetic knowledge:

(a)   treat genetic diseases;
(b)   prevent genetic diseases;
(c)   reduce the occurrence of non-genetic diseases;
(d)   enhance human traits;

(e)   promote individual and family happiness;
(f)   save resources for the society; and
(g)   respect the well-being of future generations.

Any action applying genetic knowledge is predicated on the presumptions of these dichotomies, such as nature vs. nurture, biological determinism vs. social determinism etc. With the development of human genetics and the success of the genome mapping project, more weight seems to be put on the innatism and biological determinism or genetic determinism of the balance. And the radical or hard determinist claims that all human diseases, traits, and behaviors are determined by genes, and leaves no room for the environment and the individual's free will. However, except for a few human diseases and traits, many of them are the result of interaction between multi-genes and environment. For some of them, such as mentality, the role of socio-cultural environment cannot be ignored. It is well known that a child will not develop intelligence, despite the presence of the relevant gene, if he/she is isolated from an interpersonal or social environment. Also, the genes which cause the onset of some cancers have been known, but nobody can deny the role human behavior or environmental factors play in the onset of cancer, such as smoking, radiation, and carcinogenic chemicals.

Radical genetic determinism also leads to a medical solution of social problems. If all human behaviors are predetermined by genes, then not only at risk behaviors but also unethical and illegal actions can be explained by abnormal genes and can be corrected by gene therapy, rather than by health education, moral education, or correction institutes. It will also raise some ethical and legal issues on whether offenders should be responsible or accountable for their unethical or illegal actions, because they have no choice by their free will. Genetic determinism in general, radical version in particular, will produce an over-expectation to genetic knowledge and an over-ambitious genetic program, and in turn make genetics be discredited as eugenics has been.

As for the enhancement of human traits (some colleagues include prevention of diseases in "enhancement" – I do not think it is appropriate), it raises no less issues. First, with regards to those human traits which should be enhanced, it will be a problem for which no universal agreement is foreseeable. Second, people with undesirable traits who are not treated for enhancement are most likely to be stigmatized or discriminated. Third, there is no guarantee that the enhancement of human traits will not lead to eugenic practices such as occurred in Nazi Germany.

It is acceptable to define human gene intervention or engineering as the deliberate alteration of the genetic material of living cells to prevent/treat disease or enhance traits; somatic cell gene intervention or engineering is defined as the procedures that alter the DNA of the body's differentiated cells, that is, cells that lack capacity to transmit genetic material to children; and germ-line gene intervention or engineering is what changes the DNA of reproductive cells.

Before we discuss the ethical issues in human gene intervention (or engineering) we have to make clear the criteria of how to establish whether an action is obligatory, prohibitive, or permissive. The criteria are: if the result of an action will improve the present world situation with certainty according to a certain belief system, then the action is obligatory; if the result of an action undoubtfully worsens the present world situation according to a certain belief system, then the action is prohibitive; and if the result of an action is dubious to whether it improves or worsens the present world situation according to a certain belief system, then the action is permissive. However, when we apply these criteria to gene intervention or engineering, we have to take other variables into account.

First, gene intervention or engineering for treatment is only at the stage of experiment; lots of things are uncertain. In the case of somatic cell gene intervention for treatment, we know in theory that it is very likely that the therapy can result in the cure of certain genetic diseases. However, during the therapy, we cannot categorically exclude the possibility that after the domesticated retrovirus used as a normal gene carrier has been integrated into the host's somatic cell, it will be activated and cause virus infection in the host, or activate other pathogenic factors, such as dormant cancer-causing genes, although it is not very likely on the basis of present experiences. And, so far, an animal model which will help us to know what really happens in the host after the domesticated retrovirus enters into it has not yet been successfully created. So somatic cell gene intervention for treatment is permissible and now is better used as an experimental treatment which can only be applied to a serious disease caused by the gene we exactly know and treated by no other therapy, and cannot be widely applied to many other diseases. And even when somatic cell gene intervention for treatment is permissible, therapists have the obligation to obtain informed consent from patients, go through strict review procedures before initiating therapy, and carefully to monitor patients' conditions during therapy taking the necessary precautions and safeguards.

Second, in the case of germ-line cell intervention or engineering for treatment, there are more complicated factors apart from the doubts in somatic cell gene intervention, because it involves the uncertain changes which can be transmitted to future generations, and it is not sure whether these changes will be in the best interests of our offspring and whether such changes will be consented by them if they are able to make choice by themselves. This is a new variable and raises a new ethical problem: What is our responsibility to future generations. Is it permissible for us to leave such uncertain changes in the body of our future generations? There are no sufficient reasons to permit us to do so. I agree that germ-line gene intervention for treatment is indefensible at present. But I don't agree that it should not be categorically disapproved (Edgar *et al.*, 1995), because the meaning of the phrase "categorically disapproved" is ambiguous. If it means that in the long run the germ-line cell gene intervention for treatment applied to humans will be able to become defensible one day (which I believe it will) then I agree. But if it means that although germ-line gene intervention for treatment is indefensible now, however, in certain circumstances we can try it – then I don't agree. Because at all times, we are obliged to do good with certainty to our future generations, and we are not permitted to encourage so many uncertain and perhaps harmful changes.

Third, in the case of gene intervention or engineering for enhancement, there are many more variables. First, which traits are to be enhanced? What is the criterion used to select which traits should be enhanced and which should not? One Chinese geneticist said that 1.6 m in height is the most desirable because it will save so much resources for the society. His criterion is to save resources. However, the officials in the department of sports and many coaches would say that 2.5 m is most the desirable, because it will help sportsmen and sportswomen win more gold medals in the Olympic Games. How can one resolve such a conflict? Or, is any individual permitted to enhance one of his/her traits, such as height, weight, skin or eye color, power of muscle, speed of running, and others? Does the enhancement of one trait disturb the balance within his/her body established before the enhancement, or even lead to weakening of other traits? The use of gene intervention for enhancement will be a slippery slope and lead to eugenic practices that stigmatize or discriminate against those who bear traits which were not enhanced (see Juengst, in this volume).

Germ-line intervention or engineering also means that the present generation could impose its values on future generations, because any intervention for enhancement in germ-line cells will be without the consent of future generations. Future generations who wish their ideal height to be 1.8 m will condemn us if we choose 1.6 m as the desirable height by manipulating germ-line cells. It will also open the door to massively violate the right to autonomy and the ethical principle of respect for persons. Indeed, the possible negative consequences of the manipulation of germ-line cells are unimaginable now. The damage to the human gene pool is one of these possible consequences. "Playing God" will lead to unexpected and negative consequences, which may prove difficult to remedy. There is, I believe, a moral borderline between gene intervention or engineering for treatment and for enhancement. I agree that the use of gene therapy for enhancement may become widely prohibited, and the use of germ-line gene therapy for enhancement purposes should then be categorically prohibited. But I don't agree with the "but": it should not be categorically disapproved as unethical in all imaginable circumstances (Edgar *et al.*, 1995). As I have argued, what does "categorically disapproved in all imaginable circumstances" mean? Which imaginable circumstances should be specified. Today the use of germ-line intervention or engineering for enhancement should be categorically prohibited; we should wait until a new emerging condition provides sufficient reason for us to consider it again (Agius, 1990c; Anderson, 1989; Bankowski *et al.*, 1991; Glover, 1989; Walters, 1989a; Walters, 1989b).

Institute of Philosophy
Chinese Academy of Social Sciences
Beijing, China

BIBLIOGRAPHY

Agius, E.: 1990a, 'From Individual to Collective Rights, to the Rights of Mankind: The Historical Evolution of the Subject of Human Rights', in Busuttil, S. *et al.*: *Our Responsibilities Towards Future Generations*, The Foundation for International Studies, Malta, pp. 27ff.
Agius, E.: 1990b, 'Towards a Relational Theory of Intergenerational Ethics', in Busuttil, S. *et al.*: *Our Responsibilities Towards Future Generations*, The Foundation for International Studies, Malta, 1990, pp. 73–93.
Agius, E.: 1990c, 'Germ-line Cells – Our Responsibilities for Future Generations', in Busuttil, S. *et al.*: *Our Responsibilities Towards Future Generations*, The Foundation for International Studies, Malta, 1990, pp. 133–142.
Agius, E.: 1994, 'Patenting Life: Our Responsibilities Towards Present and Future Generations',

in Agius, E. & Busuttil, S. (eds.): *What Future for Future Generations?* The Foundation for International Studies, University of Malta, 1994, pp. 99–198.

Anderson, W.F.: 1989, 'Human Gene Therapy: Scientific Considerations', in Beauchamp, T. and Walters, L.: *Contemporary Issues in Bioethics*, Belmont, California: Wadsworth, 3rd ed., pp. 513–520.

Bankowski, Z. and Capron, A.M. (eds.): 1991, *Genetics, Ethics and Human Values: Human Genome Mapping, Genetic Screening and Gene Therapy, Proceedings of the XXIVth CIOMS Round Table Conference*, CIOMS, Geneva, Switzerland, pp. 126–177.

Bell, W.: 1995, 'Why Should We Care About Future Generations?' *Future Generations Journal*, No. 15, 1–7.

Busuttil, S.: 1990, 'Preface', in Busuttil, S. *et al.: Our Responsibilities Towards Future Generations*, The Foundation for International Studies, Malta, pp. 13–15.

Confucius: *The Analyst of Confucius*, in Chan, W-T: 1964, *A Source Book in Chinese Philosophy*, Princeton, NJ: Princeton University Press.

Edgar, H. and Tursz, T.: 1995, 'Report on Human Gene Therapy', *Proceedings*, I, International Bioethics Committee of UNESCO, pp. 29–50.

Glover, J.: 1989, 'Questions About Some Uses of Genetic Engineering', in Beauchamp, T. and Walters, L. (eds.): *Contemporary Issues in Bioethics*, 3rd ed., Belmont, California: Wadsworth, pp. 525–534.

Gunatilleke, G.: 1990, 'Enhancing Social Responsibility for the Future – Some Conceptual Issues', in Busuttil, S. *et al.: Our Responsibilities Towards Future Generations*, The Foundation for International Studies, Malta, pp. 143–155.

Held, V.: 1993, *Feminist Morality: Transforming Culture, Society, and Politics*, Chicago and London: University of Chicago Press, pp. 43–63, 174–214.

Macelli, T.: 1990, 'Responsibilities to Future Generations – the Scope', in Busuttil, S. *et al.: Our Responsibilities Towards Future Generations*, The Foundation for International Studies, Malta, pp. 49–65.

Parfit, D.: 1984, *Reasons and Persons*, Oxford, U.K.: Clarendon Press, pp. 351–441.

Serracino Inglott, P.: 1990a, 'The Rights of Future Generations: Some Socio-Philosophical Considerations', in Busuttil, S. *et al.: Our Responsibilities Towards Future Generations*, The Foundation for International Studies, Malta, pp. 17–25.

Serracino Inglott, P.: 1990b, 'The Common Heritage and the Rights of Future Generations', in Busuttil, S. *et al.: Our Responsibilities Towards Future Generations*, The Foundation for International Studies, Malta, pp. 67–72.

Streeten, P.: 1990, 'Intergenerational Responsibilities', in Busuttil, S. *et al.: Our Responsibilities Towards Future Generations,* The Foundation for International Studies, Malta, pp. 157–171.

Van der Veken, J: 1994, 'Responsibilities Towards Future Generations: Philosophical Reflections', in: Agius, E. & Busuttil, S. (eds.): *What Future for Future Generations?* The Foundation for International Studies, University of Malta, pp. 157–176.

Walters, L.: 1989a, 'Genetics and Reproductive Technologies', in Veatch, R. (ed.): *Medical Ethics*, Boston, MA: Jones and Bartlett, pp. 212–228.

Walters, L.: 1989b, 'The Ethics of Human Gene Therapy', in Beauchamp, T. and Walters, L. (eds.), *Contemporary Issues in Bioethics*, Belmont, California: Wadsworth, 3rd ed., pp. 520–525.

Weiss, E.B.: 1990, 'Intergenerational Justice and International Law', in Busuttil, S. *et al.: Our Responsibilities Towards Future Generations*, The Foundation for International Studies, Malta, pp. 95–104.

ALEX E. FELICE

GUARDIANSHIP BY PEER REVIEW IN GENETIC ENGINEERING AND BIOTECHNOLOGY

I. INTRODUCTION

This volume is an uncommon opportunity for geneticists to join with other interest groups from law, philosophy, and politics and to reflect on the possible social and economic outcomes of their work as it concerns the future of humanity. For me, who comes from the background of a small unpretentious although productive and to some extent competitive research group (and from a very small though rapidly developing country) there is an added responsibility to address the interests of less powerful players in global science and technology in general, and in genetics in particular.

Thinking about future generations poses unique challenges. On the one hand, some of the interests and concerns of future generations could be considered to be much the same as those of the less powerful groups in contemporary societies. Women, children, ethnic minorities, migrants, primitive peoples, confront persistent problems of access, equity, equality, and entitlement with respect to new science and technology. On the other hand, it may be perturbing to consider the long-term outcome of an imbalance between the slow evolution of the genetic background while human life style and environment undergo very rapid alterations. I shall argue that in both contexts, the interests and concerns of future generations may be considered by widening the scope of existing peer review groups.

II. THE PROMISES AND PITFALLS OF BIOTECHNOLOGY

It is often stated that when compared with traditional science none of the targets of modern genetic engineering and biotechnology are entirely new (Pramer, 1990, p. 9). Up to the recent past, traditional or classical biotechnology, mostly based on the use of microbes in fermentation processes, has led to the production of many food types across a broad spectrum of human societies. Breeding has led to considerable gains in

E. Agius and S. Busuttil (eds.), Germ-Line Intervention and our Responsibilities to Future Generations, 117–129.

improved genetic stock and increased agricultural productivity. In the meantime, many new genetic breeds of plants and animals have been released into the wild. Gene pools of species that have survived at least three major extinctions have been altered by natural means for centuries. It is proper to ask whether there is a fundamental difference if genetic innovation is the result of mutagenesis and natural selection as exclusively hitherto, or if it is by human intervention for therapeutic or socio-economic purposes. In one sense, the past is a marvelous collection of genetic experiences from which there is much to learn. Despite eugenics, genetics has contributed much to human well being and will certainly continue to do so. Its own success gives rise to theoretical questions which range from the trivial to the profound.

Unfortunately, genetics, genetic engineering, and biotechnology have suffered from a loose terminology due to media hype that increased public expectations beyond rational ends while simultaneously giving rise occasionally to irrational fears. Nevertheless, there are recognizable and legitimate concerns that deserve attention without burdening the science with excessive regulation.

'Genetic engineering' and 'biotechnology' are largely synonymous. Both terms refer to a broad area of technology sectors in life science applications, which are led forward by gains in a few enabling sciences, including cell and molecular biology, especially recombinant DNA, molecular genetics and in biochemistry, cellular and microbial physiology. Consequently, the scope of traditional fermentation processes and breeding programmes are enhanced becoming more precisely targeted and considerably accelerated. Undoubtedly, genetic engineering and biotechnology promise to improve human health, decrease pain and suffering of sick children, control the scourge of new viral infections and cancer, put a nutritious plate on every table, turn the turbines of clean environmentally friendly power plants, and provide renewable materials for many purposes with a sound ecology. Without wanting to oversimplify the issues, few can rationally argue that these are not noble, deserving goals which will inevitably improve the human condition world-wide. The obstacles to globalization are political more than scientific or technological, and the solutions are therefore political, too.

Many professional and international organizations have invested effort in disseminating the new technologies across the globe in order to prevent them from becoming the sole purvey of a privileged few, but much remains to be done in this respect. In particular, the absorbing capacity of the less developed countries and their own commitment to the process has to increase. The UNESCO, Microbial Resource and Information Network (MIRCEN) and the series of conferences on the Global Impact of Applied Microbiology and Biotechnology (GIAM) are fora for biotechnologists from developing countries to meet with their peers from the more advanced countries. It seems important to extend participation in this kind of meeting to public opinion leaders and politicians as well.

Perhaps, the hallmarks of the new genetic engineering which distinguishes it from traditional biotechnology are precision and rapidity. The element of time intrudes to add new dimensions to integrated science management. There are two considerations to be made: The first is that projected backwards, the past experience offers much guidance in avoiding the pitfalls of biotechnology. The second is that projected into the future. We are on the one hand concerned with ethical and moral thoughts which mainly regard practice mechanisms, i.e., how do we provide the best genetics services possible in a manner compatible with human dignity. On the other hand, we have to face a larger question that results from a disparity in the timeframe of nature, which spans thousands to millions of years, compared to the human timeframe of one to two generations. The calendar of politics is even worse, rarely spanning more than a decade.

III. THE STATE OF THE ART

To discuss in detail the theoretical issues in a proper context of modern genetic engineering and biotechnology and the critical role of molecular biology and genetics, one ought briefly to review the state of the art as it can be seen from the standpoint of any rational expectations that we might have. I shall then look at the future from four perspectives, that is, the short term, the intermediate term, the long term and in particular, the very long term.

At this stage, I think that we have a fairly good understanding of the structure and organisation of a few genomes, human and others. (Human Genome, 1991–1992, *Program Report*, pp. 191–218, 1992; Lewis, p. 1, 1994.) In 1995, the entire DNA sequence of an infectious organism was published in Science (Fleischmann, 1995, p. 496), and many others will follow in due time (Fraser, 1995, p. 349). The human genome project has advanced to a position at which some investigators are convinced that large-scale sequencing is now possible. Some have started with the expectation that the whole project may now be completed well ahead of schedule, and for once within budget. After that is accomplished, then the real genome project will start. The challenge lies in understanding the function and control of the genome. Although all genes are physically present in all cells, only a fraction is functional in any single cell type or at any time during development. Furthermore, the cells of any complex organism differ in one fundamental way. There are those cells we shall call somatic cells which will die with the death of the individual and there are the germ cells, the gametes which are perpetuated by reproductive means. DNA recombination, i.e., the exchange of genetic material between two parental genomes is a physiological event that accompanies gametogenesis.

In the meantime, the genome picture emerging is that of about three billion building blocks or nucleotides, organised into 46 chains or chromosomes, two of which have to do with gender determination, the rest with psyche and soma. Among these is the code for about 100,000 genes, which, surprisingly account for less than 3% of the whole genome and of which, less than a tenth have been identified. In contrast, about 4,000 genetic disorders are cataloged (McKusick Vols. 1 & 2, 1992). Some are common, others rare. Some are due to one gene, others the result of interactions between a few or many genes called polygenic disorders. It is estimated that each one of us harbours from eight to twelve genes that if inherited from both parental lineages would cause genetic disease.

We do not know how many species each with its own genome exists in nature though we are sensitive to the need to preserve as many of these as possible. Combinatorial chemistries and rational drug design will undoubtedly relieve the pressure on biodiversity to provide us with biochemical sources for new drugs and other speciality chemicals. As a result, the use of replaceable biomaterials will increase while the exploitation of natural products from the wild will decrease.

As far as any intervention is concerned, much can be done with diagnosis, especially in regard to presymptomatic detection of carriers and couples at risk of having sick children with hereditary disease. Thus, prenatal diagnosis which contrary to many views has in fact saved the lives of many fetuses, and population testing have advanced to the stage of becoming standards of care. The same can be said for certain applications in agriculture and animal husbandry. However, efforts to actually intervene at the level of DNA have stalled beyond interventions with the simplest of genomes. Human gene therapy is not a procedure that will be common place tomorrow, nor next year. There are about 100 trials authorised for phase I/II research. They are intended to evaluate safety of the procedure rather than efficacy. The large majority are in the sphere of cancer treatment instead of genetics. The problems are technical due to inflammatory or immunological reactions to the vector molecules or in targeting the abnormal genes. Technical solutions may be anticipated perhaps in five to ten years. The technology applied to transfer genes across species is called transgenomics. It is useful to improve products such as the quality of meat and milk, for the production of biotherapeutics in Pharm-animals and perhaps to provide a large supply of organs suitable for transplantation without the hazards of rejection (xenotransplants).

Another important issue to consider is that whereas thus far there has been a consensus that gene therapy should be limited to somatic cells, a strong argument is made that germ-line therapy should be validly considered for the most serious hereditary diseases which are inevitably fatal. There also seems to be a consensus arising that whereas the actual DNA sequence should remain in the public domain as "The Common Heritage of Mankind," the resulting products and processes may be conferred with temporary protection through intellectual property rights legislation.

I do not think that there is an urgent or pressing need for new legislation. The science has a good history of being sensitive to public concerns and it needs time to develop further understanding of its own scope. There is sufficient time to engage in public debate and there are organizational tools already at our disposal that provide for human assurance in research. They have worked well in the past and can be strengthened.

IV. THE ELEMENT OF TIME

One way of looking at the theoretical or practical issues raised by the new technologies may be to consider them with respect to various degrees of a very long time span. In the short or immediate term, I include those issues connected with right to life, right to know or not to know, protection of autonomy and of privacy, discrimination and insurability. In the intermediate term, are those of patentability, insurability, germ-line gene therapy, trade displacements, and biowarfare. In the longer term, are those of possible changes in the gene pool, discrimination, biodiversity. These are not trivial issues but not insurmountable. I think that the most disconcerting issues may be those in the long to very long term. Some of these go to the heart of the problem of development. Others may be associated with maladaptation to different rates of change in human societies and the relative stability of genomes.

The intrusion of time may be considered to have been coded in DNA. Remember that the genome of any species living today is a recollection of its millennial genetic experience. Although genomes are relatively stable, the concept of a static gene pool absolutely resistant to change, which seems to underlie many concerns on genetic interventions, is wrong. Genes undergo a very slow process of natural change which may or may not result in an advantage for the organism. Subsequently, if reproductive fitness is at least undamaged, the advantage is also for the species. The advantageous genes of the remote past may then become the disease genes of today (Beet, 1946, p. 75; 1947, p. 212; Allison, 1954, pp. 290–4).

A few predominant single-gene disorders such as sickle cell disease, favism and thalassaemia or cystic fibrosis are thought to have high carrier rates today because of the selective advantage that the heterozygous carriers had in the remote past. The heterozygotes with only one copy of the gene were resistant to environmental infections; malaria in the case of the red cell disorders and perhaps cholera in the case of cystic fibrosis. In the past, those homozygotes who inherited the abnormal genes from both parents were sick and died early. With proper treatment, many today survive and have satisfactory lifestyles. However, for every homozygote treated it can be calculated that there are thousands of healthy heterozygous carriers who are not candidates for any genetic manipulation. Consequently, it can be seen that it is most unlikely that any genetic intervention on sick homozygotes can have any effect on the prevalence of any gene even in the very long term.

Certain polygenic disorders such as diabetes and obesity or mental illness may be considered to fall in this category too. When food was, for long periods of evolutionary time scarce and consequently much energy had to be expended in search of food, genomes evolved to conserve calories and essential nutrients. Today, that over a remarkably short period of recent time, food has become abundant, and lifestyle sedentary, our genome finds itself deficient or incompetent and imbalanced, with metabolic disease and perhaps also mental disorder frequent. In this context, the gravest concern for the future may be with respect to the genetics of human behaviour.

Allow me to quote the infamous Unabomber who writes: "I attribute the social and psychological problems of modern society to the fact that society requires people to live under conditions radically different from those under which the race evolved" (In Wright, 1995). I now speculate: it seems as if the genomes of today and likely of the future may not be properly attuned to the psycho-social demands of technologically induced and rapid changes in lifestyle. Many studies mainly involving twin comparisons indicate that the contours of the human mind, our emotions, wants, and needs are nearly as much the result of heredity as of learning experience. The intrinsic or genetically determined patterns of human behavior may be dominated by the remote imperatives to survive, to reproduce and to transmit one's own genes into posterity. It is fair to surmise that a future civilized society will have a markedly different set of imperatives. It is difficult to predict what these might be, but they may among others include the psychosocial domain which one might temporarily call an informative or communicative genetics. Evolutionary psychology is a new field of science that, as described in a recent article from *Time* magazine: "examines the mismatch between our genetic makeup and the modern world looking for the source of our pervasive source of discomfort" (Wright, 1995). While the major promise of genetic engineering and biotechnology may be in enhancing the socio-economic condition of mankind, i.e., the physical dignity, the major pitfall may well be in the resulting contrast between the "Ancestral Environment" which is imprinted in the genome, and the "Modern or Future Environment." That is to say, that human dignity may be endangered by a growing gap between the genome and the environment or lifestyle. One may wonder whether over a long period of time, genomes might adapt to the information overload or if there could be "behaviour genes" with which to manipulate adaptation.

I may add that is unlikely that genomes will continue to evolve so as to grossly modify human anatomy or physiology. It is quite possible that the present-day structure of man is close to a perfect balance with the environment. It is unlikely that homo futuris will have grossly different arms, legs, or skull, or a grossly different cellular metabolism. It is more likely that the future course of human evolution will involve the mind and interpersonal communications. Information technology, by enabling new forms of nearly instantaneous communications between distant humans may, as much as genetic engineering and biotechnology, mould mental evolution in an as yet unpredictable manner. The genetics of mind and the nascent field of evolutionary psychogenetics are poorly understood and lag markedly behind our understanding of somatic genetics. They are good candidates for heavy investment of research resources because it is likely that the boundaries of knowledge may advance rapidly if well founded research is supported. Already, a number of brain-specific gene clones have been isolated, and a brain-specific gene map may soon be available. However, as has been said with respect to the whole genome studies, understanding of function and genetic control is limited. Gene control and gene-environment interactions may have much to bear on the psycho-social situation confronting future generations. Clearly, a technology of such import cannot remain the restricted domain of a few privileged groups or countries and globalization has to improve.

V. ASSURANCE BY PEER REVIEW

The question arises then as to what mechanisms do we have available to ensure that human interests are protected and good science receives the resources it needs to thrive, while limited resources are not wasted on poor or unproductive science. Development requires the mobilization of public and private resources while addressing the legitimate concerns of the less powerful among us including those of future generations. Legitimate concerns have to be coupled with legitimate responsibilities. It is futile to argue that uncontrolled genetic engineering and biotechnology will widen the gap between the rich and the poor, or further debase the disadvantaged without proposing routes to enable less developed countries to participate in the global development process. Progress is not possible without a commitment from the developing countries themselves to provide at least a basic infrastructure for science and technology. Too often this is lacking. Once again, the problems are in the realm of politics and once again that is where the solutions ought to lie.

The new genetic engineering and biotechnology adds new dimensions to integrated management procedures, ones projected long forward versus the present. Science itself thrives on foresight. Adventure, whether it is in the search for food or for mates, for development gains or intellectual satisfaction is one of the most valid components of human civilisation. Experimentation is an essential step in development of individuals as much as of entire societies, and novelty begets progress. The history of the human species has been one of continuous genetic innovation. One may ask again if it is fundamentally different if genetic innovation is the result of mutagenesis and natural selection or if it has been by human intervention for therapeutic or socio-economic purposes. A type of decision such as this falls beyond the limits of any single investigator scientist or research and development group. Rather, these should be in the domain of collective judgment based on education and experience in the process of peer review.

It can be seen then, that the element of time is an essential ingredient in passing judgment on the value of science and technology and in particular of genetic engineering and biotechnology for human development. A review of past and present experiences provides guidelines in making projections for the future. At the global level, it has been proposed and widely discussed, I must add not without many reservations that a Guardian might be appointed to promote or protect the genetics interests and genetics' concerns of future generations. Lately, it is further proposed that the Trusteeship Council of the United National Organizations may be revitalized to play the role (De Marco, G. and Bartolo M.).

One may legitimately, then, ask how does organized society determine what priorities or even what criteria are to be employed in establishing priorities for the allocation of limited human and other resources to ensure that as much as possible, good science is done and bad science is not. It is necessary to proceed both with confidence and caution and to seek a wide and educated public consensus.

Invariably, whenever scientists write research proposals for submission to mainstream funding agencies, we are asked to address the impact that our research might have on patient care or relevant spheres of application. All too often, perhaps, this question is approached rather trivially. In the sector of human genetics research, any individual scientist or research group firmly sustains the view that their proposed research will lead to progress in the diagnosis or care of a major disease. Cancer, AIDS, Aging, or other conditions common or mainly afflicting women, children or minorities are often indicated in this section of a research proposal.

Many investigators feel that in the intense competition for the limited research resources existing today, their chances of success are increased if they indicated that the proposed research could benefit a strong lobby or an "attractive" category of people. Research scientists have good cause for concern. Last year, less than ten percent of research proposals submitted to the U.S. National Institutes of Health were approved and funded. Research support in Europe is even less than this. Excluding Japan, the rest of the world does not have a review mechanism that could have provided comparable data. To be fair, my impression is that this section of funding applications plays a very small role in influencing the final judgment of the awarding bodies. But, perhaps, as I shall argue, the review system, has here an opportunity to give more weight to the special interests and concerns of unique interest groups, which should include among the others mentioned above also those of future generations. I am not too much concerned here with the activities of rogue scientists who break the rules. The system has proven robust enough in enforcing integrity in scientific research.

In mainstream science, funding agencies ask for the views of expert "Peer Review Groups" to whom research proposals are referred for evaluation and rated by priority for funding.

The peer review groups seek specific answers on the quality of the proposed research, the clarity of the question asked, the relevance of the proposed work to the state of the art in the field, the significance of the work proposed, the appropriateness of the research method, the competence and the experience of the investigators, the protection of human participation in the research, the likelihood that the proposed research may advance the frontiers of knowledge or improve the human condition. Peer review groups are occasionally asked to consider specific areas such as AIDS research, Gene Therapy, Human Genome or others and make recommendations for the development of designated programs followed by a call for proposals. Although the subject of much criticisms, peer review groups have proven themselves as one of the most effective mechanisms that promote the development of strong and competitive research and development programs. The only tried alternative is a command system that too often depended on patronage.

Science has many unwritten rules which young investigators learn by apprenticeship, usually in the course of post-doctoral training with an experienced mentor. One of these is that the collective wisdom and the good judgment of a number of experienced scientists exercised together with lay advice in peer review groups has served well to direct resources into productive research while guaranteeing human assurance. The conduct of peer review assumes that a consensus can be reached on questions of scientific research. They serve to identify those lines of research that are mature for major steps forward or of others which although less developed are begging for well thought consideration or of social and economic needs that could benefit from intense attention. It is commendable that so many scientists and lay participants give so much of their time to sustain the peer review system. They do so with enthusiasm convinced of their long-term value. Their contribution in speaking also for future generations can be further enhanced by including larger numbers of the lay educated public.

The wider scope for public participation in peer review groups requires increased efforts to educate the publics concerned. The effort should start in the schools by including the principles of genetic engineering and biotechnology in the curricula of high schools and junior colleges. Other publics which act as intermediaries in popular information such as journalists and politicians need to be targeted too. Good educational programs such as the ones promoted by the University of Reading and UNESCO are available for the purpose. International professional organizations such as the American and the European Societies of Human Genetics or the Human Genome Organisation ought to undertake the added responsibilities of providing for public education, developing guidelines, creating pluralistic consensus, and supporting external peer review groups in small or less developed countries.

There are strong grounds on which to argue that peer review groups are effective tools that we already have and with which to integrate the public viewpoint and the potential interests or concerns of future generations into the research enterprise of today. There is much to learn on the conduct of peer review and it has been the topic of a number of studies. [An updated bibliography is appended to the text.]

VI. CONCLUSION

My vision, then, is one of intense support for the further development of genetic engineering and biotechnology on a global basis. At the national, local, or institutional level, widened peer review groups provide assurance that good science is adequately and stably supported while guarding the position of the powerless and future generations. At the global level, a guardian, perhaps the Trusteeship Council of the U.N., might function to integrate and sustain the efforts of local and institutional peer review groups.

Department of Pathology
School of Medicine
University of Malta

ACKNOWLEDGEMENTS

I thank my wife, Joanna, who is on the staff of the University of Malta Library and my student assistant, Ms. Stephanie Bezzina Wettinger, for their research and editorial assistance.

NOTE

[1] Effective peer review requires commitment and participation from a number of participants both within and outside the actual research and development enterprise. It is likely that in practice, different institutions will find different mechanisms to best employ local resources and share experience with similar groups from other countries. At the University of Malta, in addition to the formal research ethics committee of the Faculty of Medicine, which is empowered to judge research applications we have two other informal groups, which although lacking any regulatory powers, meet openly to discuss the implementation of innovative or advanced technologies in health care practice. One group is based in our laboratory and addresses specifically the best conditions, technical and ethical, under which new procedures from Molecular Biology or Genetics may or may not be introduced into practice. The second group is based in the Department of Theology and undertakes ethical and philosophical review of standards of care in genetics. It is anticipated that the second group will in time publish a record of its proceedings.

BIBLIOGRAPHY

Allison, A.C.: 1954, 'Protection Afforded by Sickle-Cell Trait Against Subtertian Malarial Infection', *British Medical Journal* **1**, 290–294.

Beet, E.A.: 1946, 'Sickle Cell Disease in the Balovale District of Northern Rhodesia', *East African Medical Journal* **23**, 75–86.

Beet, E.A.: 1947, 'Sickle Cell Disease in Northern Rhodesia', *East African Medical Journal* **24**, 212–222.

De Marco, G. and Bartolo, M.: 1995, *A Second Generation United Nations for Peace in Freedom in the 21st Century*, reviewed by M. Hogg in *The Sunday Times*, October 29, pp. 5–6.

Fleischmann, R.D. *et al.*: 1995, *Science* **269**, 496.

Fraser, C. M., *et al.*: 1995, 'The Minimal Gene Complement of Mycoplasma genitalium', *Science* **270**, 349.

Human Genome: 1992, 1991–92 *Program Report*, pp. 191–218. U.S. Department of Energy, Washington, D.C.

Lewis, R.: 1994, 'French Team Completes Physical Map of Human Genome', *Genetic Engineering News* **14**:1, 1.

McKusick, V. A., 1992, *Mendelian Inheritance in Man*, Vols. 1 & 2, John Hopkins University Press, Baltimore, MD and London, UK.

Pramer, D.: 1990, *Biotechnology: Promise and Pitfalls*, The Chinese University Press, Hong Kong, China.
Wright, R.: 1995, 'The Evolution of Despair', *Time* **146**:9, 32–38.

BIBLIOGRAPHY ON PEER REVIEW

Grol, R. and Lawrence, M.: 1995, *Quality Improvement by Peer Review*, Oxford University Press, Oxford, UK.
Daniel, H. D.: 1994, *Guardians of Science: Fairness and Reliability of Peer Review*: V C H Publishers.
Sturkie, J. and Phillips, M.: 1994, *Peer Helping Training Course,* Resource Publications.
Research Ethics, Manuscript Review, and Journal Quality: 1992, *Symposium on the Peer Review – Editing Process*, San Antonio, Texas, 1990. American Society of Agronomy.
Lock, S.: 1991, *Difficult Balance: Editorial Peer Review in Medicine*, British Medical Association.
Chubin, D.E. and Hackett, E.J.: 1990, *Peerless Science: Peer Review and U.S. Science Policy*, State University of New York Press.
Iverson, C. (ed.): 1989, *American Medical Association Manual of Style*, 8th ed.
Pearson, A.: 1987, *Nursing Quality Measurement: Quality Assurance Methods for Peer Review*, Wiley & Sons Ltd, UK.
Harnad, S.R.: 1983, *Peer Commentary on Peer Review: A Case Study in Scientific Quality Control*, Cambridge University Press, UK.

DAVID HEYD

ARE WE OUR DESCENDANTS' KEEPERS?

The allegory which the mystics tell us – that we men are put in a sort of guard post, from which one must not release oneself or run away – seems to me to be a high doctrine with difficult implications. All the same, Cebes, I believe that this much is true, that the gods are our keepers, and we men are one of their possessions.

Plato (*Phaedo* 62b)

I. THE POWER OF METAPHORS

In one of the earliest uses of the metaphor of guardianship, Socrates tries to convince his interlocutors that taking one's own life is wrong since life itself is not one's "possession" but rather belongs to the gods. Further-more, human beings have a positive duty to guard life, not to let it go. The gods are our keepers in the sense of ownership, but we are their delegates or trustees in the role of guardians of the property they own, namely life. Running away from this duty is a violation of trust. Typi-cally, Plato, the rationalist with the deep sense of the mystical, appeals to the Orphic tradition and to the allegorical rendering of the idea of re-sponsibility for one's own life. He describes the doctrine as "high" (which could be interpreted as both "noble" and "mysterious") and cautions us of its difficult implications.

Twenty four centuries later, a Maltese government proposal appears under the title "A 'Guardian' for Future Generations":

Time is now ripe enough to give a practical substance and a concrete form to our responsibilities towards future generations. In order to pass from theory to praxis and from mere words to action, a specific mechanism (The Guardian) should be developed to encourage our re-sponsibility towards generations yet to be born and, at the same time, to safeguard their interests from present threats (Agius and Busuttil, 1994, p. 315).

Here the Platonic "allegory" is given a significant shift: from responsibil-ity towards oneself to responsibility to others, from the individual to an

E. Agius and S. Busuttil (eds.), Germ-Line Intervention and our Responsibilities to Future Generations, 131–145.
© 1998 *Kluwer Academic Publishers. Printed in Great Britain.*

unidentified collective, from the present to the long-term future, from mere physical life to the overall conditions and welfare of the human race. However, in line with Plato's awareness of the problematic "allegorical" character of the idea of guardianship, the Maltese proposal puts the word "Guardian" within inverted commas. My purpose in this article is to examine critically the idea of guardianship and particularly its applicability to the highly controversial issue of our responsibility to the human genetic pool. My thesis will be that the idea of a genetic Guardian (in the sense of the Maltese proposal) is, to use Plato's words, "a high doctrine with difficult implications."

Unlike poetry, in philosophy metaphors are a mixed blessing. Their suggestive power often stands in direct proportion to their misleading effect. Guardian, in the above mentioned context of the responsibility to future generations, is a "twice-removed metaphor": it rests on an older, legal, term, which is itself a metaphor of the literal sense. Thus, from the parent guarding the child playing in the park, we derive the idea of a court-appointed guardian of an orphaned child; and from this legalistic idea a further concept of a guardian of future children is extrapolated. My aim is to challenge this latter metaphor and to expose the misleading analogies which underlie it. It should be pointed out that the issue is not semantic, that is a question of the appropriate use of words in certain contexts. My concern is typically philosophical, namely an analysis of the logical and metaphysical conditions under which the use of the metaphor "guardian" makes sense.

As an aside I should point to a strikingly similar double shift of metaphorical meaning, which could lead to an analogous philosophical fallacy, viz. the term "heritage." Originally it refers to "anything that is or maybe inherited" (*O.E.D.*); from this literal, biological, meaning the metaphorical meaning of something which is "worthy of preservation" is derived; it is usually associated with past human endeavour and achievement (typically, works of culture); but from that, we get a third use, "genetic heritage", which is a blend of the biological and the cultural; this, as I will try to show, surreptitiously leads us to the controversial conclusion that the genetic pool is a valuable achievement which is worthy of preservation.

Going back to the metaphor of guardianship, a preliminary remark should be made on the context of its use. In the original Maltese proposal, the general context is primarily environmental. It relates to the "unprecedented power to change the environment extensively, lastingly, and in part irreversibly." Indeed, it also mentions the "manipulation of genetic material" which could "affect the future of all life on earth" (Agius and Busuttil, 1994, p. 313). But the document also refers to the threat to genetic diversity, and considers on an equal footing the depletion of natural resources and the genetic intervention in the nature of the human species. My purpose in the following sections will be to cast doubt on the analogy between the two issues, and particularly to make the claim that the idea of guardianship cannot be extended from the environmental context so as to apply to the genetic. An illuminating metaphor becomes philosophically misleading in precisely this apparently innocuous shift. In other words, although the legal metaphor of guardianship retains the essential logical structure of the term in its literal origin, the metaphysical use (either in its Platonic, self-regarding sense, or in its inter-generational sense) is problematic, or even fallacious.

II. GUARDIANS AND TRUSTEES

Moral concern for human beings is a universal requirement. But concern for the weak is the specific point of a moral system, since dealing with the more powerful is naturally constrained by considerations of self-interest. The weak are typically those who are dependent on us, those who cannot always make a claim to, let alone insist on getting, what is due to them. In the more extreme cases, the role of morality is to ensure that their rights and interests are represented, that is, to put forward their claims in their stead. Thus, minors, retarded people, the incompetent and the politically weak are those who need someone who will speak on their behalf, who will serve as their advocate in the world of claims and counter-claims.

There are two ways in which such interests can be represented: guardianship and trusteeship. Guardianship is a "relationship arising from the natural incapacities of infants and persons of unsound mind, and sometimes other categories of persons, to manage their own affairs". A trustee is "a person having nominal title to some right or property which he holds not for his own sole beneficial interest but for the interest of another or others" (Walker, 1980). Both concepts seem to constitute a binary relation (the guardian and the guarded, the truster and the trustee); but actually they are based on a tertiary relation: one person (the court) nominates another (the guardian) to protect the interests of a third party (an orphan); or, in the second case, one party (shareholders) appoints another party

(the board of directors) to take care of its property (investment). Though
the purpose of both legal titles is the same, viz. the representation of the
interests of another person, the "direction" of the relationship is different.
The responsibility of the guardian is to the ward, rather than to the court;
the responsibility of the trustee is to the truster, rather than to the object of
his care.

This distinction between the two kinds of tertiary relations, which
prima facie look quite similar, is not a legal nicety. In the context of our
responsibility to future generations, two different sources of a moral duty
can be discerned, as we shall see, on the basis of these two kinds of legal
metaphor: on the one hand, a duty towards the beneficiary party (the
guarded) often created by a third party (the court); on the other, a duty
towards the party creating the duty (the truster) the object of which is a
third party (or some property). Hence the grammatical difference between
the active (guardian) and the passive (trustee) roles of the "middle" term
in the three-place relations. In other words, our duties towards future
generations can be seen under two perspectives: we are either guardians
of future generations (or the planet) or trustees of God on behalf of future
generations (or the planet).

The object of guardianship is usually a human being, whereas that of
trusteeship is non-human property.[1] Thus it is natural to speak in the
ethics of inter-generational responsibility of guarding the interests of
future people, but of maintaining the integrity of the environment as it
was given to us in trust. But the two cannot be easily separated, first,
since care of the environment is a major aspect of our duty to future
people, and secondly, since the duty to future people is often derived
from the power delegated to us by the creator to act as procreators, that is
as his trustees in this world. Thus, the first commandment in the Bible,
"Be fruitful and multiply," is the epitome of the relations of both guardianship and trust: human beings are expected to exercise their procreative
power so as to continue and extend God's creation.

The philosophical issue in our context is, therefore, the following: can
the human gene pool be treated either as a weak party which cannot
manage its own affairs and hence needs a guardian; and, alternatively, can
it be seen as a property of God over which we are entrusted with a duty of
care. Again, the first perspective emphasizes the status of weakness and
dependence of the guarded, while the second focuses on the wishes of the
truster, both being sources for the respective duties of care. We shall
examine these two perspectives in turn.

III. DEPENDENCE VS. WEAKNESS

Future people are in the most basic sense dependent on us, living in the present. Their very existence is a function of our choice. So is their number. To these two forms of inter-generational dependence, which have characterised human procreation since time immemorial, there is now added a third form: that of identity. Through modern techniques of genetic manipulation and engineering the present generation gains far-reaching power over the personality traits and biological properties of future people. Furthermore, these dependence relations are not mitigated by any form of potential retaliation or retribution, since, at least when we are speaking of long-term population and genetic policies, the alleged perpetrators of the harm are not going to be around when its victims come to life. As has been noted again and again in the literature on in-ter-generational justice, the very position of future people further down in the temporal stream makes them dependent on the decisions of those located in the upper part of the stream (like the analogical dependence of residents on the lower reaches of the river on the responsible behaviour of factory owners along the upper reaches).

This asymmetrical relationship between present and future people, which constitutes the latter's dependence on the former, is the fundamen-tal reason for the inapplicability of contractarian theories of justice to the inter-generational sphere. Since Hume, it has been acknowledged that the so-called "circumstances of justice" involve the condition of mutuality, that is the potential benefit (and harm) which can be caused by each of the partners in the system of justice. The fact that we can benefit and harm future people but cannot be benefited or harmed by them is a major obstacle in subjecting our responsibilities towards future generations to the principles of justice. John Rawls and others have tried to overcome these obstacles by extending the circumstances of justice to a hypothetical situation, a theoretical construction in which temporal location does not play a role, and thus to evade the issue of non-mutuality. My argument is that even if such a construction can solve the question of what Rawls has called "just savings" (that is the conservation of resources and capital), it cannot guide us in issues of genetic responsibility.

The basic reason for my argument is that dependence, in the above mentioned wide sense, does not imply weakness. Dependence of the created on the creator is distinct from the dependence of the weak on the powerful. The conflation of the two senses of dependence is misleading.

Hence, future people cannot be seen as weak subjects requiring protection through guardianship. The fact that they are born without choosing to be so, and particularly with properties they have not chosen for themselves, is a logical necessity which does not grant them any rights against us, nor indeed determine any criteria which might serve their alleged Guardian in deciding how to protect them. Their lack of independent identity, that is identity which is prior to our choice, equally undermines both the attempt to extend the principles of justice to genetic choices as well as the idea of nominating a Guardian for the genetic welfare of future people.

Let me explain. Future people may be considered under two logically distinct categories: actual and possible. Actual people are those who are not around yet, but whose existence and identity is (or will be) fixed independently of our choice. Possible (or potential) people are those whose existence and identity is a matter of our choice, that is to say, not only are they not in existence now, but the question whether they will exist and with which properties is a matter of our decision. The child of our next door neighbour who is going to be conceived and born in the future is (at least for us) an actual child, as are the millions of children who are going to be born in India next year. The children of a couple which has the means of family planning, as well as the children to be born of a project in genetic engineering, are according to my definition, possible people.[2]

Now, for logical reasons possible people cannot be, in my opinion, moral subjects, nor can they have moral rights. One simple argument supporting this proposition is that rights are derived from interests, and interests are a function of what the subject of the interest is, that is to say, his desires, needs, life-plans, sources of possible satisfaction, etc. A possible person is, by definition, someone whose desires and interests are undetermined, since her basic biological or psychological features are not fixed. How can such a person have rights and claims on us, or even be a subject of utilitarian, welfarist consideration? The inability to manage one's own affairs occurs either in the case of possible persons, where there are no subjects and hence no "identifiable affairs", or in the case of actual people when there are subjects but who lack the capacity to manage their affairs. Only in the second case can we coherently speak of a guardian. Contrary to the rhetoric of the Maltese Proposal, possible people cannot be described as "mute," except for the trivial sense that non-existing entities cannot speak! (Mute people are standardly understood as those who have something to say but are unable to do so for some reason or other.)

Another way to put this is by exposing the limits of the counterfactual argument often used in the moral discourse of rights: had you acted more responsibly, I would have been better off now. The problem is that in contexts where the action in question is the very birth or the determination of one's genetic make-up the counterfactual situation means non-existence, or being someone else. It is, accordingly not I who would have been better off under the counterfactual claim (for instance, had my parents not conceived me with my genetic defect). It is equally not I who could be the source of moral claim on any other party for having created me as I am.

The legal concept of a guardian refers to a mechanism through which the best interests of another party are represented. The guardian of a person is expected to promote his or her "best interests" (e.g., in cases of minors), or sometimes to act on the basis of what the guarded person would have done, had she or he been competent to do so (e.g., in cases of comatose patients). But the best interests of future people is a highly indeterminate notion. First because we do not always know what these interests will actually be, and secondly because the idea of genetic engineering is exactly the capacity to create people with different properties and interests than our own. The guarded in the case of future people are not "given" people, but potential beings who are molded by us. Therefore, the constraints on the way we mold them cannot logically be derived from what they are, from their so called "best interests," nor indeed from the alleged weakness in their being unable to express what they wish to be.[3]

The harsh conclusion of the preceding paragraph should be mitigated by the admission that as a matter of fact we know much about what future people are going to be like, that is to say, the distinction between the actual and the possible as defined above is relative, and rarely neat in its application. We know, for example, that human beings in the future will be such that their best interests would include clean air or moderate climate. On this basis we can claim that radioactivity in the atmosphere or global warming are dangers that we should prevent for their sake. However, genetic engineering may cause a change so radical in the make-up of the human organism that very basic systems of human needs and interests may be modified. Science fiction provides us with endless constructions in which a newly created human species does not need and cannot even appreciate what we need and value for ourselves. A striking example is Frank Herbert's science fiction heros, called the Optimen, whose biological perfection lies precisely in their being on the one hand immortal, yet on the other hand incapable of (and completely uninterested in) procreating (Herbert, 1968).[4] Could we argue against the creation of such creatures in the name of their interests?

IV. FROM INDIVIDUAL TO IMPERSONAL VALUES

But the inability to anchor the idea of guardianship in terms of the weakness of future individuals leaves room for supporting this kind of care for future generations in non-individual terms. Even in the original legal sense of guardianship, the objects of care can be tradition, values, knowledge, scientific heritage, standards of truth, that is to say impersonal entities or values. Now, some of the justification for constraining genetic engineering, and particularly germ-line genetic manipulation, appeals to such impersonal values. Thus, for instance, the Maltese Proposal speaks of the "genetic heritage of posterity" or "the unity of the human species." These are surely not the interests of individual future human beings, but trans-individual values which are irreducible to individual welfare. These values are analogous to the values of the integrity of the planet, the variety of biological species, the conservation of the natural resources. But while the impersonal values both of cultural heritage and of environmental integrity can at least be indirectly justified in their effect on individual human beings and their interests, the value of the unity of the human species, or the diversity of the genetic pool are values which determine what human welfare is and hence cannot be justified by it.

The value of the existing genetic pool is often appealed to in the discussion of the ethics of biotechnology. But at a closer look it proves to be difficult to justify. There is the aesthetic dimension, namely the value of variety and diversity. There are also considerations of genetic utility (or the health of the human species) associated with such diversity. Evolution itself cannot be expected to yield higher forms of living without the underlying variety of the gene pool. Then there is the conservativist view which holds that there is a value in preserving the human species as it is, and warns against the emergence of a completely different species which would replace homo sapiens. But these are all arguments which cannot serve to justify the idea of a Guardian, since guardians by definition represent the interest of human beings. That is to say, they will guard certain impersonal values and heritage only for the sake of future people. But if our remote descendants are going to be different from us, maybe they will have no interest in or appreciation for these impersonal values, including the variety of the genetic pool or "the unity of the human species."

Again, as suggested in the previous section, as a matter of fact human beings, even after radical genetic intervention, are going to be quite similar to us (like the similarity of our genetically engineered tomatoes to the original natural stuff we remember from our childhood ...), and hence can be presumed to have basically the same interests as we do. However, this does not undermine the general theoretical proposition I am making here; moreover, the further into the long-term future we go, the less we can be sure of this similarity, thus making our guardianship responsibility at most relatively short-term.

It should also be remembered that what we are, that is human nature, is also a function of a long process of evolution, typically a random process with no pre-ordained direction or goal, guiding principle or purpose. It is true that the kind of beings we are dictates the way we form our values, but this does not mean that the way we are is, from the point of view of the overall evolutionary process, the best or the optimal. And even if there is a sense in which evolution is a form of progress, there is no reason why human technological intervention should not serve to accelerate it, for instance by directly removing "bad" genes.

Genetic guardianship can take as its object three different kinds of impersonal value. The first, which may be called Platonic, believes that the aim of genetic planning is the perfection of the governing castes. This is a monolithic or substantialist conception in arguing for an *a priori telos* or essence of human nature. Nietzsche holds a version of the Platonic eugenic vision. Secondly, there is the Kantian attempt, typically formulated in a recent book by Kurt Bayertz, to ground genetic responsibility in a kind of meta-value, an uncontested principle which does not presuppose any substantialist value (Platonic perfection, God's image, rationality). Bayertz suggests autonomy as the only moral basis for genetic policies, since the very ability to make reasoned choices of values and ideals is the essence of human nature and the ultimate value which must be preserved (Bayertz, 1994, pt. III).[5] Thirdly, there is what I referred to as the conservativist view, that is to say the "sanctification" of the existing situation, the diversity and variety of the human genome. This approach, in a sense a Millean conception, hails heterogeneity and richness as a quasiaesthetic value, with some utilitarian overtones.

The first, substantialist view is hardly helpful for the creation of a guardianship policy in genetic planning, since there is no chance of agreement on the essence and goals of humanity, and any attempt to implement such a policy would be treated in our liberal society as typically totalitarian. The second view, which appears typically liberal, is, I think, in the last analysis a substantialist view. It is true that most of us think very highly of autonomy, but there is no logical inconsistency in imagining future people whose psychological make-up will be such that autonomy will not play any important role in their life. And one should just think of the long periods in human history in which the culture of autonomy was completely absent to see the parochiality of our modern insistence on the uniqueness of autonomy. The conservativist conception has already been criticised as parochial in another sense, that is, as taking what we are as the standard and norm for what any human beings should be.

However, beyond the problem of specifying the object of genetic guardianship, if indeed none of these three conceptions can withstand criticism, there is the question of the principles of control which would guide the genetic Guardian. Again, one can think of three options.

The first, eugenic interventionism, which would recommend an active intervention in human procreation so as to improve, perfect and optimize human beings and the human gene pool. This is Plato's vision of the role of his utopian rulers, not accidentally also called "Guardians"!

The second kind of guiding principle is the active preservation of the present state, an approach which tries to maintain the basic features of the genetic pool by active intervention so as to correct the deterioration of the pool caused by man-made environmental damage or medical progress. (Note that the ability to save the lives of defective children leads to the long-term deterioration of the gene pool).

The third approach is passive non-interventionism, which is usually associated with the moral prohibition against playing God, or the sanctification of the natural (over the artificial, technological, or scientific). A genetic guardian in this case has the sole role of blocking any kind of germ-line genetic engineering and securing the free development of the human species either towards some sort of a telos or by the random vicissitudes of Darwinian evolution.

We have so far seen that founding the principles of responsibility in genetic policies on the idea of guardianship, is highly problematic, both if we take future individuals as the object of the guardianship or if we take our guide from impersonal values. While in the first case the problem is logical, in the second it is metaphysical. However, responsibility to future generations can adopt what we referred to earlier as the opposite "direction," that is, we owe a duty of genetic care neither to future people nor to impersonal values but to some kind of third party.

V. TRUSTEES OR STEWARDS?

In his Second Treatise of Civil Government, John Locke follows Plato in prohibiting suicide (as well as the murder of others) in terms of what may be called the "trusteeship argument." We all serve God who created us and assigned us to the role of doing his will. Being God's property we have no authority to take any life. Locke precedes Kant in forbidding human beings to treat each other (and themselves) as means rather than ends. But unlike Kant, the reasoning behind this prohibition is theological: our life, both as individuals and as "one community of nature," is given to us as trustees, since we are "sent into the world by his order." Man is responsible to another authority (God) "to preserve himself and not to quit his station wilfully" (Locke, 1967, sec. 6).[6] The same trusteeship argument is applied to non-human property, that is to all natural resources, which were given to us by God for our use and not for their annihilation or waste (sec. 31).[7]

Can Locke's argument be extended so as to cover future people, or the more abstract value of the human gene pool? First, obviously the argument presupposes a series of theological beliefs: that there is a God, that God created human beings, that God empowered them to continue the work of creation by procreation, and that this procreation should not take forms which would lead to changing the nature of the human species. Furthermore, one should be able to specify the ways of genetic engineering or eugenic policies which would count as "spoiling and destroying" what God gave to us. The merit of these beliefs will not of course be discussed here, but they are not all universally shared, and hence cannot serve as a firm basis for an ethics of genetic policies. Although I believe that the biblical commandment to be fruitful and multiply should be understood as a delegation of divine creative power to human beings (who become able to create life almost ex nihilo), it is difficult to derive from it a limitation on the freedom to (pro)create people of different nature than ours. In other words, if "the image of God" is understood as the very power of creating new life, rather than a fixed nature or essence of human beings, then it cannot be considered as violated by genetic engineering.

And once theological explanation is abandoned in favor of Darwinian evolution, the trusteeship argument becomes vacuous, since it is hard to identify a possible "truster" to whom we are responsible. Theoretically one could think of our ancestors as those to whom we owe the duty of trust. It is as if they entrusted us with the genetic heritage as they did with the cultural and scientific heritage. Like God delegating procreative power to Adam and Eve, so do our parents create (and raise) us so as to give "them" grandchildren. This argument may sound convincing but only to a certain point. Surely our immediate ancestors have an interest (which we may have a duty to respect) to continue the human species, and in the way they wish it to be. But the further in the future we go, the less we have such an interest. Psychologically, human interest in the nature of future people extends to three or at most four generations (corresponding to the natural human life span and generational overlap).

Furthermore, the content of the trust, even if we can specify the "truster," is difficult to articulate. What are the trusters' exact wishes? Is it the human genome as it is, which ought to be preserved? Is it the human power of self-creation and improvement? Is it such eugenic potential which is the object of the trust, but only within certain bounds or constraints? [8] And what are these limits? Unless the voice of the truster (divine or human) is actually heard, we have no practical guideline to serve the proposed Guardian of future generations. And I believe this is not just a theological problem. Between the two limits of the spectrum, the complete ban on any intervention (including that which removes human illness) and the Frankenstein-like wild genetic molding of a new species, we ourselves have at most only vague ideas about those limits. If we cannot formulate the conditions of trust as trusters, how can we hope to be able to guide our policies on the assumption that we are the trustees?

The difficulty in defining the Guardian's necessary traits of character is directly related to the difficulty of describing the object of the interests which he or she is expected to protect. The Maltese Proposal tries to do exactly that:

> The appointee to the office would need to be an eminent person, without known prejudices, and having practical wisdom, integrity, modera-

tion and humility, with an ability to feel the pain and share the joy of people who will live at a great distance from us in time (Agius and Busuttil, 1994, p. 316).[9]

I am not sure there can be such a person. The nature of the joy and pain of future generations might be so different from ours that we should be extremely careful in trying to act on them as the basis of our genetic policies. Our close ancestors in the previous century could not have visualized many of the pains and joys of the modern, post-industrial age; how can one expect a Guardian to be able to share the basic psychological responses of people who are going to live in two thousand years' time? Maybe humility should lead to the precisely opposite conclusion, the abandonment of any attempt to represent the interests of future generations. Furthermore, no Guardian can be free from prejudice in the relevant, genetic sense. The Guardian, being herself of the present genetic make-up, is a person with a bias towards widely shared values like autonomy, variety, rationality, creativity, intimacy, etc. But with a little effort one may imagine creatures who would not have such preferences. Is there not a danger that the Guardian rather than protecting future people's interests would become a paternalistic authority, exercizing his power by temporal remote control?

VI. CONCLUSION

I have criticized the idea that responsibility to future generations can be grounded in the concept of representation. My attempt was to show that the interests of future generations, particularly on the level of their genetic makeup, cannot be represented, either by a guardian or by a trustee, and that genetic responsibility cannot be analysed in terms of either the rights of future individuals, or the impersonal value of the "integrity" of the human gene pool. The terms in which the critique is developed are both logical and metaphysical. This brings us to the conclusion that the extension of the legal metaphor of guardianship to inter-generational care is potentially misleading, either giving rise to incoherent consequences, or making metaphysically controversial and vague assumptions.

But to end on a positive and more hopeful tone, (to quote once more the Platonic text of the *Phaedo*) for guardianship of future generations to be a "high doctrine", it must be emphasized that my critique by no means implies that there are no alternative grounds for our responsibility to future generations in general and the care we ought to take in genetic policies in particular. The fundamental justification of and constraints on genetic engineering lies in what we, actual present people, wish to project into the future, in what we, in the deep sense, want our descendants to be like. Most of us want the human species to continue; moreover, we want to create people who are basically similar to us. These considerations, though they are not of a moral nature in the sense of the protection of the interests and rights of others, are serious and noble enough to serve as guiding principles in genetic planning.[10]

Furthermore, we do not have to nominate a guardian who will be uniquely sensitive to the pains and joys of future people, since we know that we are going to live with the next generation or two. We usually hope that our children and grandchildren will be sufficiently similar to us, due to the generational overlap. But this obviously implies a "time discount" in which the more distant people are in the future, the less concerned we are with their nature and character. This is not merely a psychological fact but a conceptual limit on the criteria of the permitted and the forbidden in genetic planning.

The ethical (or rather meta-ethical) lesson of this long argument is that although we definitely are our brothers' and childrens' keepers, we cannot be our remote descendants' keepers.

Department of Philosophy
Hebrew University of Jerusalem,
Israel.

NOTES

[1] A notable exception is Stone (1987, pp. 44–5), who argues that non-persons like inanimate natural entities have legal standing and can be the object of guardianship. In his theory of guardianship, the status of "distant people" looks, accordingly, like an easy case. Stone mentions the idea of a guardian as a means to protect biological species in general (p. 74).

[2] I elaborate extensively on this conceptual distinction and its moral ramifications in my book (Heyd, 1992, chapters. 4 and 6). The distinction which looks somewhat counter-intuitive is given meaning in the context of a full moral theory of procreation, which I refer to as "genethics."

[3] My conclusion stands in direct opposition to the Maltese Proposal which says that "[i]n this respect – the ability to appraise what one's interests are and how they might best be protected – future generations are similar to those that our society has declared legally incompetent" (Agius and Busuttil, 1994, p. 315).

[4] I am grateful to Uriel Heyd for turning my attention to Herbert's science fiction book and to

its deep relevance to the philosophical problem of defining the "optimal" human being (for the logical reasons to which I am referring).

5 Bayertz, who believes in "autoevolution," seems to leave no room for a guardian of future generations, since each generation is contributing to the self-evolving humanity. But this approach does not seem to be consistent with his argument for limiting eugenics. However, Bayertz is fully aware of the problems of guardianship for future people in his section on Advocatory Ethics (pp. 249 – 251).

6 Note the metaphor "quit the station" which is identical to Plato's "guard post" from which we should not "run away" in the above quoted section from the *Phaedo*.

7 Note that Locke is implicitly making the same distinction as made here between the relation of trusteeship and that of guardianship. The latter is described in detail in sections 59–60 in the context of the duty of care of the incompetent and the authority of parents over children which is derived from it.

8 John Hospers suggests a model of man's responsibility to nature which is based on cooperation and perfection (rather than despotic exploitaiton on the one hand and divinely grounded stewardship or trusteeship). Nature is treated as potentially perfectible, and man is cooperating with the actualisation of this potential. Hospers' conception might be applied also to the human genome which, after all, is an integral part of nature (Hospers, 1974, pp. 32–40).

9 The text also addresses the modalities of the functioning of the Guardian, the scope and limit of his institutional authority. These raise interesting issues in political philosophy and the way duties to future generations should be distributed within the present generation (e.g., should these be dealt with on the national or the international level?). Despite their importance, I cannot deal with these issues here.

10 John Harris discusses the restrictions on genetic engineering and the responsibility to future generations, but does so in terms of our obligations rather than "their" rights. Although I do not agree with some of his arguments, I find it significant that there is no role for a guardian in his theory (Harris, 1992, chapter 8).

BIBLIOGRAPHY

Agius, E. and Busuttil, S. (eds.): 1994, 'A Guardian for Future Generations' in *What Future for Future Generations?*, The Foundation for International Studies, Malta, pp. 313–17.

Bayertz, K.: 1994, *GenEthics* (trans. S.L. Kirkby), Cambridge University Press, Cambridge, UK.

Harris. J.: 1992, *Wonderwoman and Superman*, Oxford University Press, Oxford, UK.

Herbert, F.: 1968, *The Eyes of Heisenberg*, Sphere Books, London, UK.

Heyd, D.: 1992, *Genethics*, University of California Press, Berkeley, CA.

Locke, J.: 1967, *Second Treatise of Civil Government* (ed., P. Laslett) Cambridge University Press, Cambridge, UK.

Plato: 1963, '*Phaedo*', in *The Collected Dialogues* (trans. E. Hamilton and H. Cairns), Princeton University Press, Princeton, NJ.

Stone, C.: 1987, *Earth and Other Ethics*, Harper and Row, New York, NY.

Walker, D. M.: 1980, *The Oxford Companion to Law*, Clarendon Press, Oxford, UK.

STUART F. SPICKER

THE UNKNOWABLE EFFECTS OF GENETIC INTERVENTIONS ON FUTURE GENERATIONS (OR, WHO GUARDS THE GENETIC ENGINEERS IN DEMOCRATIC REPUBLICS?)[1]

"Is genetic engineering the promise of a Golden Age?
Or does it presage an apocalypse?"
> – Dr. Michel Salomon

"Neither one...."

> – Dr. Jean Bernard
> (Salomon, 1983, p. 283)

INTRODUCTION

Rather recent efforts to affect the human germ-line by means of gene therapy Suzuki and Knudtson describe as a transition from the study of "mortal soma" to "immortal germ plasm" (1989, p. 203; Heyd, 1992; Hubbard and Wald, 1993, pp. 113–116). Only mentioning this scientific process, I turn to the political philosophy and mechanisms that have proven efficacious when permitting and funding genetic engineers, who live and work in democratic republics, not only to acquire and possess but also to control the use of knowledge acquired by germ-line engineering that bears critically on the lives not only of temporally contiguous but of temporally distant unborn generations. I refer to *Human Dignity and Genetic Heritage* (a study paper published by the Law Reform Commission of Canada in 1991) observing that as an example of "bureaucratic bioethics" it does not adequately address the bearing of the results of genetic engineering on future generations. I then introduce the notion of accountability and argue for its importance in guiding decisionmakers in democracies which rely on approval by the people. I conclude that since there is neither a fixed human nature nor human genome, that geneticists be held accountable to others – who must exercise democratic safeguards – as these scientists conduct their research by adhering to the prevailing normative standards of scientific inquiry.

E. Agius and S. Busuttil (eds.), Germ-Line Intervention and our Responsibilities to Future Generations, 147–163.

I. FROM EUGENICS TO TRAIT ENHANCEMENT

Although it is tempting to address the eschatological suggestion raised by Suzuki and Knudtson's notion of "immortal germ plasm," suffice to say that present interventions to eliminate genetic "flaws" or "defects" should not be occasion for one to presume that these various "flaws" have only long-term and irreversible effects; indeed, the effects may, in the end, be no more than transient, i.e., affecting transient qualities. Thus, interventions to "correct" these qualities by manipulating the DNA of externally fertilized, very early human embryos (Agius, 1989; Bonnicksen, 1994) continue to raise controversial scientific issues among geneticists.

I do not here intend to explore these and other important scientific issues, however. Rather, I shall attend to the political philosophy and mechanisms that have thus far been proven efficacious when permitting, encouraging, and funding geneticists who live and work within democratic republics (in this context the genetic engineers) not only to acquire, possess and, in all likelihood, overestimate their knowledge (Hubbard and Wald, 1993, p. 115), but to control and use the knowledge (Bondeson *et al.*, 1982) that issues from their research for "projecting where we are going" (Lappé, 1978, p. 84). I shall assume that this research is quite likely to affect the lives not only of temporally "contiguous" but of temporally quite "distant" unborn generations – keeping in mind of course that "it is impossible for the gene alone to be all determining" (Lappé, 1978, p. 93).

In *Out of the Night: A Biologist's View of the Future*, published in 1935, as well as in his address, *Man's Future Birthright*, delivered at the University of New Hampshire on November 21, 1957, the geneticist, Hermann J. Muller, is principally preoccupied with the exciting beginning of a new era regarding future technological possibilities, not only the new reproductive practices he witnesses (what we today describe as a variety of modes of collaborative reproduction and of procreating the generations to come), and the "peril of overpopulation" (Muller, 1935; 1958)[2] but especially "the possibility of positive biological improvement of mankind" (1935, p. viii), what Muller calls "a real biological upbuilding" (1935, p. 10). He says, faithful to the Darwinian standpoint:

> ... man's heritage of genes remains the basis that has made possible the development and retention of this learned culture. And his genes have attained their high capabilities only as a result of the cruel struggle for existence, that time and again awarded the spoils to the most

proficient, while the devil took the hindmost except when he had a friend. For under the unrelieved population pressure that exists in nature, there can seldom be enough to go around (1958, pp. 3–4).

To be sure, Muller is writing at a time when the notion of eugenics "has become a hopelessly perverted movement," since the term had come to represent a justification of Nazi atrocities and perversions and for imposing far more than "some slight limitation on the numbers of the most grossly defective"[3] (1935, p. ix). Indeed, in the closing pages of *Out of the Night*, Muller remarks:

> Mankind has a right to the best genes attainable, as well as to the best environment, and eventually our children will blame us for our dereliction if we have deliberately failed to take the necessary steps for providing them with the best that was available, squandering their rightful heritage only to feed our heedless egotism (1935, pp. 113–114).

As for the question whether there is adequate moral or theological (Archer, 1994) justification for geneticists or the state[4] to intervene in human germ cells and to attempt therapeutically to enhance certain heritable traits deemed desirable (Fletcher, 1974, p. xvi), or even to eliminate by so-called "negative eugenics" undesirable or even harmful traits (Golding, 1978; Suzuki and Knudtson, 1989, pp. 205–206), the general consensus is that this is premature, especially if one must first determine the goals of an entire society, itself a complex philosophical and political undertaking. Indeed, a number of geneticists as well as the lay public worry that geneticists' attempts at enhancement could turn out to be medically hazardous, i.e., the risk of causing harm and a further "restricted life" (Archer, 1994, p. 232; Kavka, 1982, p. 105) could exceed the potential benefits, especially since geneticists are not at all confident that they could repair or "correct" damage already done. In addition, such interventions may prove "morally precarious," since societies are not at present equipped to make the moral decisions (and subsequent public policy) required, and could not be relied upon to control discriminatory practices that are virtually inevitable as a result of enhancement engineering. Moreover, so far there is "no straightforward and unequivocal answer to this question of moral justification, since," as Ernest Nagel points out, "every answer presupposes some moral theory and social philosophy; and there is no moral system that all reflective men accept as uniquely correct" (Nagel, 1978, p. 95), nor can, on my view, one appeal to self-evident principles or some theory of natural rights for justification even if such a moral theory or a theory of natural rights or a compelling social philosophy was to surface. However, there is, as we shall see, a method for determining consensus with respect to public policy other than by a self-appointed scientific élite.

II. AN EXAMPLE OF BUREAUCRATIC BIOETHICS

It is interesting immediately to note that in a bi-lingual Canadian study paper – *Human Dignity and Genetic Heritage* – the chapter "Towards Genetic Justice" addresses, but only briefly, "the clinical or research context, where ultimately the policies will be expressed" (Knoppers, 1991, p. 59). The principal author, Bartha M. Knoppers, a health lawyer, advances the thesis that two new [moral?] "founding principles" must be introduced in this context: (1) reciprocity (exchange) and (2) mutuality (civic responsibility).

'Reciprocity' recognizes the "inequality between knowledge held by ordinary individuals and that held by practitioners of medical genetics." The text goes on: "Justice requires that such knowledge be redistributed in a way that is beneficial to the less well-informed, that is, the ordinary citizen. A redistribution is essential to ensure that knowledge of medical genetics is not used by the state to impose decisions on individuals and to monitor their compliance" (Knoppers, 1991, p. 69). Unfortunately, the report never addresses the fact that the process of educating and informing the public has substantial costs that a society may not be willing to expend because these costs are unrealistic, although quite rational, and the public's behavior therefore is unlikely to change. The consequence of such an attitude is that it leads all of us to remain ignorant of many things.

The report concludes that this exchange "is best conducted within the physician-patient relationship, which has traditionally provided protection for the patient." From this conclusion (itself highly dubious, I might add) it is clear that the "vision" of the Study ends with living patients, their families, and perhaps "society at large"; it does not, however, adequately address the bearing of the results of genetic engineering on "distant" contingent future generations. Simply stated, the Law Reform Commission of Canada limited itself to the potential risks of physical and psychological harm to living Canadians, who may consent to undergo or to reject "genetic tests" proposed by their physicians and/or genetic counselors. In fairness, however, one should point out that the Commission does recognize that a tacit moral issue underlies the "transfer of knowledge from the molecular biologist or geneticist to the general practitioner" (p. 70) and, in time, to the community; here the Study hints that some sort of "civic responsibility" exists, though the Commission does not articulate it as a responsibility of geneticists to future generations. Moreover, this responsibility leads directly to the second principle – mutuality – or the sharing of genetic information.

'Mutuality' arguably assumes that individuals have a social obligation not to withhold information useful to other members of their families, where to do so could cause harm. Indeed, should the individual refuse to share this information under Canadian law, his or her physician would be legally authorized to override that refusal. Secondly, the principle of mutuality presumes that each individual has a community or civic obligation as well, one that may find expression when a citizen voluntarily participates in such programs as the banking of DNA for future use. Finally, at one point the Study asserts: "Genetic disease implicates not only the individual but also the family, the community, and *contingent future generations*" (p. 72, my italics). No more is said concerning future generations, however, nor is the term even included in the Glossary.

III. "PROTECTING" FUTURE GENERATIONS: A LOGICAL WORRY

Before we consider the paradox in claiming to "protect" future generations, we should humbly iterate a very recent lesson: we are frequently ill prepared to live through what was only portended some years ago, and to discover now that things have not turned out as we expected. Indeed, what better example of this "surprise" than the recently challenged assumption that many scientists accepted as true and now turns out to be false: namely, that the incidence of tuberculosis-related deaths would remain low, without resurgence, so long as the appropriate and heretofore potent medical remedies continue to be taken as prescribed. In short, virtually no one was prepared for the recent and rapid rise of multidrug-resistant strains of the causal agent, *Mycobacterium tuberculosis,* that is inefficacious in 14% of cases treated (Lee, Klietmann and Ferrano, 1995, p. 28) that now presents us with a very dangerous situation, when not long ago we accepted the scientists' forecast that complete eradication of this disease would be achieved in the United States by the end of the

20th century. Some years from now will today's exhibited confidence in germ-line interventions be viewed in retrospect as unwarranted?

Perhaps more instructive is the continuous, though gradual, "readjustment" of our social norms: recall that not so long ago genetic counseling was valued as a new means of providing couples with the freedom to choose (with greater likelihood of intended outcome) to avoid bearing a child who would live a restricted life. Today, the same knowledgeable couples are considered "socially irresponsible" if they elect to bear such a child (Callahan, 1973, p. 6).

These empirical matters quickly lose the spotlight when we turn toward the logical worry hiding in the claim that we ought to work to protect future generations. What can possibly be meant by "protecting" future generations when protection here presumes existing persons, not imagined, individual future persons. That is, can we coherently compare our need for protection with the needs of non-existent, contingent future individuals or generations? Such a quandary is analogous to a court's considering compensation on the basis of a claim to "wrongful life," i.e., that it was wrong knowingly for a newborn's parents to conceive him (to bring him into existence), when they understood beforehand that there was a high probability that he would be born with a "predictably appalling" quality of life (Heyd, 1992, Chap. 1; Parfit, 1982, p. 148). In order to entertain the very notion of providing compensation to such a newborn, his attorney must be able to compare his client's present condition and quality of life with his previous condition and quality of life. However, courts find it incoherent to be asked to compensate a plaintiff on the basis of a comparison of the plaintiff's present condition or restricted life with his "previous" non-existence, though courts frequently compare life with death. And so, too, with claims about future possible lives when we, the living, are asked to compare them with ourselves. Thus far, I see no way out of this paradox concerning future individuals (Kavka, 1982; Parfit, 1982). Here I agree with Parfit, whom I paraphrase: unlike never existing, starting to exist should be classed with ceasing to exist, for both happen to actual people, whereas classing never existing with starting to exist (as some claim we should, because if one had not started to exist one would never have existed) does not permit a comparison of two states of existence (Parfit, 1982, p. 136). Although future generations cannot coherently claim that their predecessors (earlier generations) failed in their duty to "protect" them, they can, however, coherently claim that the governments of previous generations acted irresponsibly in failing to regulate the actions of genetic engineers, whose research outcomes and discoveries concerning the human genome could be used either to the detriment or benefit of actual future individuals/generations – this, precisely because temporally "contiguous" generations from our own, i.e., the next one or two (Fletcher, 1974) have necessarily to be affected (at least have had to be known gene carriers) before temporally distant or remote future generations could be so affected. But this scenario, I argue, is qualitatively different from claims about the present generation's duties to protect future generations. Put another way, if the actions or choices of today's geneticists are a necessary part of the cause of the existence and identity of particular people of contiguous (as well as distant) generations, then they are morally responsible not only for the known but, more importantly, for the uncertain, unknown and perhaps empirically unknowable effects of their actions on future generations (Golding, 1978, p. 511).

IV. "GUARDING" THE GENETICISTS IN DEMOCRATIC POLITIES

As early as 1973, Daniel Callahan introduced his notion of "setting limits," indicating that prudence requires us to establish "a system of prohibitions, denials and interdictions which establishes the limits of technological aggressiveness, hopes and mandates" (p. 5). By 1991, Bernard D. Davis, scientist and advocate of minimum social restraints on the basic scientific enterprise, published his edited volume, *The Genetic Revolution*, in which a number of his contributors sought to assuage the lay public's concern regarding the apparent "risks associated with deliberate releases of genetically engineered organisms" or "biohazards" into the environment (1991, p. 57; Hubbard, 1993, p. 127).

A few years earlier, Suzuki and Knudtson concluded, with respect to their discussion of "The Future of Gene Therapy": "...in the future, if gene therapy techniques are ever aimed at human germ cells, the cost of our ignorance of genetic processes would instantly become unacceptably high..." (1989, p. 207). This view is shared not only by Suzuki, Knudtson, and other members of the scientific community, but by a number of philosophers, the most prominent being the late Hans Jonas. In reading his complex *The Imperative of Responsibility* (1984, pp. 38–44) and other essays on the same theme, for example, "Technology and Responsibility" (Partridge, 1981, pp. 23–36), one is struck by his thesis, that "the survival of the [human] species is more than a prudential duty of its present members" (Partridge, 1981, p. 28). Here is not the place to outline the detailed steps of his complex line of argument, but simply to note that his argument begins with the claim that the archetype of all responsible action is "powerfully implanted in us by nature or at least in the childbearing part of humanity" (Jonas, 1984, p. 38). I note that Jonas makes no mention of the fact that many people have no interest whatsoever in bearing progeny, and, moreover, there is no justification to warrant pejorative judgments directed at them by those who do not share their values.

From this starting point, Jonas seeks to reach the conclusion that the present generation has a duty to ensure a future, that duty being, in Jonas' own words, hard to prove, since it is "not the counterpart of another's right" (p. 40). Jonas, therefore, takes another approach: He says that "Knowledge ... becomes a prime duty ... and must be commensurate with the causal scale of our action." Most importantly, he adds,

> Recognition of ignorance becomes the obverse of the duty to know and thus part of the ethics which must govern the ever more necessary self-policing of our out-sized might. No previous ethics had to consider the global condition of human life and the far-off future, even existence, of the race (Partridge, 1981, p. 29).

For Jonas, then, the distant future, being indefinite, requires imperatives and at least "self-policing" of a new sort. In the end, this will lead to the formulation of public, perhaps even global policy. If he is right, Jonas' reflections compel us to readdress the meaning of basic genetic science and the remote effects and influence or, as Jonas puts it, the "quasi utopian powers" of biomedical scientists and geneticists in democratic republics, where ignorance is itself a reason for responsible action, since ignorance can only be second best to wisdom. Jonas, furthermore, claims that "representative government"[5] is insufficient to meet the new demands on its normal principles and mechanisms (Partridge, 1981, p. 35; Brown and Wahlke, 1971). Here we can ask: is this really the case with regard to "guarding" the geneticists in democratic republics? Are parliaments, congresses, juridical processes, public agencies, and public voices (Watson and Juengst, 1992, p. xviii) inadequate to the task of monitoring accountability with respect to the power of present-day scientists pursuing their inquiries? Moreover, is it not less important who makes certain decisions than that these persons can be held accountable to the public for their actions? Can not elected bodies oversee future interests, i.e., the work of geneticists that bears on future generations? Jonas on my view is simply incorrect when he says that future generations are not represented on our present scales. Indeed, we often behave as if our actions affect future generations. Therefore, I raise rhetorically Jonas' question anew: "What force shall represent the future in the present?" And does this question itself reflect the need for a novum? Perhaps not. Let me explain.

'Democracy' is first the name for a political method or institutional arrangement a nation uses for arriving at political decisions, in which individuals acquire the power to decide by means of a competitive struggle for the people's vote.

The term 'democratic' as I employ it, then, following Joseph A. Schumpeter, does not mean government by the people. As he says, "Equating 'making decision' to 'ruling', we might define democracy as Rule by the People" (Schumpeter, 1942, p. 243). But apart from isolated referenda, the people do not make policy. They do not know or care enough to do so, nor can a workable system of tens of millions of policy makers even be imagined. Rather, through the mechanism of regular and free elections, they are able to select the most important policy makers and to hold them accountable for their pubic acts. More precisely, democracy means government approved by the people, who have the opportunity of accepting or refusing the men and women who are to rule them. Consent, then, refers not to an event in the antediluvian past when the republic was founded, but instead to a continuing process. Those in power seek the people's consent to govern, and those in opposition seek to wrest it from them. This competition ruled by self interest – politicians want power and voters want benefits – drives the system. And because self-interest, unlike, say, wisdom or virtue, exists in such great abundance, the system is feasible, workable, in a word, practical. Put another way, "the will of the people is the product and not the motive power of the political process" (p. 263).

V. CONCLUSION: THE TRUE GUARDIANS IN DEMOCRATIC
REPUBLICS

... I am persuaded that it is time we ceased pretending that we know the
correct way to arrange the life of man.
                                        – Joseph Margolis (1975, p. vii)

If a society considers the influence that geneticists can in practice exert
on the human progeny of future generations, then it must concede the
normative advantage of defining 'influence' as either enhancing future
generations, or intruding on or even endangering future generations.
Either way, one cannot avoid the expression of one ideology
(Schumpeter, 1942, p. 266)[6] taking precedence over the other, notwithstanding the fact that in both cases one is concerned with human prudential interests, even if in this case they can bear powerfully on the distant
unborn. The question is whether we can, without resorting to the category
of the sacred (the category most thoroughly destroyed by the scientific
enlightenment) have a standpoint or ethic from which we may cope with
the extreme powers which geneticists possess today, consequently increase, and are virtually compelled to use (Partridge, 1981, p. 36). The
point we must not ignore is, that it is quite likely that no prospect of moral
objectivity remains. Put more positively (though one may not celebrate
this conclusion, either), we must leave it to future generations, one at a
time so to speak, to determine their own courses of moral action, since for
us the "best actions" on the part of those who will live in the future are
necessarily unforeseen, unknowable, and in many cases spontaneous –
with the exception of course that we can be "certain," in the most general
sense, of the environmental necessities that make life as such possible.
But our concern here is not about the future environment, but rather how
to come to accept the fact that we must continue to tolerate a wide variety
of co-existent norms – appreciating the fact that any normative view
betrays its own normative loyalties – discovering in the process

> that we are unable to argue for the reasoned direction of human life as
> such. Philosophy, having not yet proffered a demonstrably valid mo-
> rality (a state of affairs disappointing to a number of philosophers but
> not, perhaps, to ordinary men and women), "must yield to ideology"
> (Margolis, 1975, p. 159) or perhaps ideologies, especially since hu-
> manity is, in our day, *Homo faber*. It seems we are actors and reshapers
> with no unarguable or definitive fixed human nature that we have thus

far been able to disclose or discover. Indeed, it may be time to dispel the myth that there is a fixed human genome, since the genetic base of what is human continually undergoes alteration as a consequence of the process of random mutation (Boone, 1988, pp. 10–11).

Perhaps I should apologize, in the end, for having reduced my view to a matter of prudence (an Aristotelian virtue to be sure); for in holding the power permanently to influence the progeny of "distant" future generations, geneticists can, in principle, seriously affect these future generations and must therefore be accountable to others.

Today, some argue, we do not possess the genetic knowledge to justify a moral right to plan for future generations, and so we should not do so, while others like Ernest Nagel cogently argue the contrary, though he claims that it is not clear that we have a moral right to implement such plans (p. 94). Thus, if we are prudent, minimum rational constraints on geneticists to assure their accountability for their actions (Pitkin, 1967, p. 11) should be exerted via democratic political processes (employing what Margolis calls "partisan adjustment"), though reminding ourselves to keep these constraints in the form of legal sanctions to a minimum (Fletcher, 1974, p. 24) as generations of geneticists proceed to explore now and in the future the innermost workings of human biology. What we can demand, however, is that geneticists stay true to what Jonas calls "the internal norms of scientific inquiry" – to remain careful in their work, keep faith with the standards of science, acknowledge the value of truth, remain dedicated, persistent, disciplined, and open-minded and, finally, share the results of their work not only with the scientific community but with the lay public as well.

EPILOGUE

In 1586, at age twenty-two, returning to his family in Florence, Galileo Galilei applied himself to the study of mathematics, particularly the work of Archimedes and what has come to be called 'specific gravity'. He soon published his first work, *La Bilancetta* [*The Little Balance*], in Italian, not Latin, the language of the Roman Church and the intellectuals; he even offered public lectures on this instrument, also in Italian, the language of the day, not Latin. It is not a radical speculation to suggest from this fact, as some scholars have, that Galileo (though not typically inclined to exaggerated modesty) was quite willing to address the Italian public, especially artisans, who neither spoke nor read Latin, but who could well build and use this instrument.

Over four hundred years have passed since that remarkable event – a scientist writing and speaking the language of hoi polloi – and thus one might have thought that we today would have learned its lesson: that genetic engineers – clearly members of a prestigious intellectual minority, empowered by the tools of experimental inquiry (McNeill, 1993) to discover and acquire, as well as possess and possibly use this new knowledge and set of techniques to affect future generations – ought not any longer be permitted to restrict their communications in democratic polities by the use of the technical language of genetic biological science merely in communication with specialists and committees of specialists within the international scientific community (Fletcher, 1974, p. xvii). That is, they are no longer warranted to remain self governed, self regulated, and the sole possessors of this new knowledge. Furthermore, as the late Joseph Fletcher observed, "... when René Dubos says that we must have 'criticism of science formulated by enlightened nonscientists' the real problem is one of getting the nonscientists enlightened" (pp. xvii, 21–22).

I therefore suggest that a relatively new task for members of legislatures, national parliaments, congresses, and juridical bodies, our duly elected guardians, is not to appoint and authorize representatives (Guardians) to participate in debates conducted under the auspices of the United Nations or other international bodies, for that matter, but rather to urge the true guardians – our elected officials – to continue to work cooperatively, yet critically, through the sanctioned democratic process, and to mandate that genetic engineers be obliged to educate those who work in the media of the long-term implications and estimated risks and benefits of their scientific work. This is, in my view, an extremely important, even urgent, need, though it is not in itself novel. Indeed, we should remind ourselves of the Jeffersonian dictum: in the end the people are wiser than any single individual can be; or Lincoln's: the impossibility of "fooling all the people all the time."

Since as we have learned war ought not be left to the generals alone, nor biology to the biologists alone (Fletcher, 1974, p. 25), the results of the work of a relatively small number of genetic engineers ought at the very least be communicated to the educated lay public, as Francis Crick suggests, by a cadre of scientifically educated journalists (Zimmerman, 1984, p. v), especially since people unhappily "know far more about their cars than they do about themselves" (Fletcher, 1974, p. 38). This scientific knowledge must not be left for translation to the public only by genetic counselors, family physicians, and clinical researchers[7] who counsel, conduct clinical research, and write in the language of the day. Furthermore, geneticists usually will admit, in their more honest moments at least, that most of the time the future bearing and risks and benefits of their work is, figuratively speaking, Greek (Latin ?) to them....[8]

> Should man disappear from the face of the earth he would not be the first species to do so.
>
> – John Passmore (1974, p. 131).

ACKNOWLEDGEMENT

I take this opportunity to thank my friend and scholar, Thomas Halper, Ph.D., for his extremely helpful suggestions following his careful and critical reading of the manuscript.

DEDICATION

To my granddaughter, Kimberly Rose Spicker – born July 10, 1995 – a joy to her parents, Merrill and Aaron Spicker, and renewed hope for the next generation.

Massachusetts College of Pharmacy
and Allied Health Sciences
Boston, Massachusetts, U.S.A.

NOTES

[1] To remain with my topic I shall not refer to the edited volumes and anthologies that proliferate madly on population control (Sikora and Barry, 1978) – the natural tendency of a population to increase faster than the means of subsistence – as discussed in writings from T. R. Malthus to H. J. Muller. Nor shall I address genetic testing, screening, and preconception counselling and what they presume or include: research on the "underpinnings of disease states and their inheritance" (Lappé, 1978, p. 84) and diagnoses of people who carry single but detectable copies of genes that (in double doses) may be said to cause "genetic" diseases; nor the molecular mechanisms of genetic diversity; susceptibility and carrier testing; individual risk prediction as a consequence of genetic manipulation (Andrews *et al.* 1994; Hubbard and Wald, 1993, pp. 36–38; Nagel, 1978); DNA banking; the patenting of laboratory

generated new-life forms (Fletcher, 1974, pp. 99–118); and the grounds of ownership or financial advantage of possessing genetic knowledge acquired by genetic engineers (Hubbard and Wald, 1993, pp. 124–126). These and related topics have already received serious if not fully adequate attention in the extant literature. I shall not even discuss our obligations, if any, to future generations with respect to (1) promoting their well-being through our reproductive choices or the quality of their environment in terms of our resource conservation (Agius, 1994; Sikora and Barry, 1978, pp. 3–13), or (2) securing for them the availability of what we judge will be needed by these still unborn descendants of whose environment we can predict very little. Finally, although an important consideration, I shall not address the moral implications of the impossibility of future generations to affect or act reciprocally toward the present generation.

2   Muller, it appears, was one of the first to draw attention to the problem of "overpopulation." The Australian philosopher, John Passmore, prefers the morally neutral term 'multiplication' to the normative (or sometimes pejorative) 'overpopulation'; for the use of the latter, he says, implies that "people are pollution" and reflects "the anti-human bias of so many ecologists" (Passmore, 1974, p. 127). Over twenty years ago, Passmore understood, that "The precise importance of population growth as a source of ecological destruction is highly debatable" (p. 128); indeed, the debate continues today, though there is now a growing consensus among the experts that the steady increase in the number of living human beings is not in itself the principal cause of ecological destruction. What we have witnessed of course, is the "continuous and increasing tempo and acceleration of population growth rather than the obscurantist and rhetorically more effective 'population explosion'" (p. 131*). (See Parfit, 1984, Ch. 17).

3   Muller has in mind what he calls "the genetics of the new society, freed of the traditions of caste, of slavery, and of colonialism." This notion of eugenics is, on his account, the "true eugenics" (Muller, 1935, p. 120). Here it should be pointed out that Muller, although acknowledging that "great men" did not owe their greatness "entirely to genetic causes," claims that humanity has the obligation to predetermine by selection of maternal protoplasm and selecting among human male germ cells, "the characteristics of some transcendently estimable man, without either of the [genetic] parents concerned ever having come in contact with or even seen each other..." (p. 111). In short, if we continue to reproduce on the basis of personal love, says Muller, "you degrade the germ plasm of the future generations" (p. 112). Such nonsense!

4   Much is now being written on these topics, including estimates of the cost of such ventures as well as whether "genetic repair" or enhancement should or could be prohibited (Gardner, 1995) given that most traits are polygenic.
     I also omit here a discussion of the perplexing problem of what, if anything, the state ought to do to regulate or override the freedom of its citizens with respect to reproductive choices germane to the procreation of future progeny, notwithstanding the daily advice provided by genetic counselors and physicians.

5   To provide a full treatment of the notions of (1) the contract between the living, the dead, and the yet to be born, and (2) public accountability in the context of representative government, one must carefully review Edmund Burke's (1729–1797) theory of representation within his more broadly cast defense of the art of political organization. He says, for example: "... the individual is foolish; the multitude, for the moment, is foolish, when they act without deliberation; but the Species is wise, and when time is given to it, as a Species it always acts right. Thus, the Burkean view of the state is that of "a partnership between those who are living, those who are dead, and those who are yet to be born." Furthermore, as to

representation, we should distinguish between the delegate (who follows the perceived voters' will, not his own), the trustee (who follows his own best judgment), and the partisan (who follows the party). As a mater of practice, in technical subjects like genetics, legislators will often take their cues from trusted experts. In addition, we should rely on institutions which have stood the test of long experience. [See, Eulau, J.C. Wahlke, W. Buchanan, and L.C. Ferguson in "The Role of the Representative" (Brown and Wahlke, 1971, p. 212).]

6   By 'ideology' I mean a useful, collective, political illusion (that may be neutral if it encompasses two or more sides constituting an ideological struggle) that reflects the psychology of a given age; not always directed to action, though it may be directed to the realization of a given objective, typically through concerted group action. I therefore distinguish ideologies (where one may triumph over another at any given historical moment) from a well reasoned justification of conduct or action, though this alone does not guarantee that we shall ever discover a demonstrably valid morality (Engelhardt, 1996, Intro. Note 3, pp. 17–19).

7   In fairness, it should be mentioned that a few educational programs on human genetics and genome analysis have been conducted for non-geneticists and the public (including editors, writers, congressional and science museum staff, lawyers, physicians, medical ethicists, representatives of state governments, and genetic support groups) at Cold Spring Harbor Laboratory, Long Island, New York, in 1992, under the leadership of Jan Witkowski, and the Director and Assistant Director of the DNA Learning Center, David Micklos, and Mark Bloom, respectively. It remains unclear whether this program addressed the ethical issues engendered by genetic engineering, the regulation of geneticists in democratic republics, or our obligations, if any, towards future generations. [See *Human Genome News*, January 1993, 4(5), p. 10.] Lori B. Andrews, *et al.*, note that such educational workshops, though designed for manufacturers, have been organized under the auspices of the FDA on the critical aspects of genetic testing technology, viewed as investigational devices (1994, pp. 142–143).

8   It is of some interest to note that notwithstanding the fact that numerous Maltese buses continue to post the aphorism – "*Verbum Dei Caro Factum Est*" [The Word of God Has Been Made Flesh] – none of the Maltese public I queried over a nine-day period could meaningfully translate the Latin.

BIBLIOGRAPHY

Agius, E.: 1989, 'Germ-line Cells – Our Responsibilities for Future Generations', in D. Mieth and J. Puhier (eds.) *Ethics in the Natural Sciences*, T. & T. Clark, Ltd., Edinburgh, Scotland, pp. 105–115; reprinted in Busuttil *et al.*, 1990, pp. 133–142.

Agius, E., Busuttil, S. (eds.): 1994, *What Future for Future Generations?, A Programme of UNESCO and the International Environment Institute*, Foundation for International Studies, Valletta, Malta.

Andrews, L.B., Fullarton, J.E., Holtzman, N.A., Motulsky, A.G. (eds.): 1994, *Assessing Genetic Risks: Implications for Health and Social Policy*, National Academy Press, Washington, D.C.

Archer, L.: 1994, 'Gene Therapy in Catholic Thought', *Journal International de Bioéthique* 5(3), 229–233.

Bondeson, W.B., Engelhardt, H.T., Spicker, S.F., White J.M. (eds.): 1982, *New Knowledge in the Biomedical Sciences: Some Moral Implications of its Acquisition, Possession, and Use,*

D. Reidel Publishing Co., Dordrecht/Holland, Boston/USA, London/England.

Bonnicksen, A.L.: 1994, 'National and International Approaches to Human Germ-Line Gene Therapy', *Politics and Life Sciences* **13**(1), 39–49.

Boone, C.K.: 1988, 'Bad Axioms in Genetic Engineering', *Hastings Center Report* **18**(4), 9–13.

Brown, B.E., Wahlke, J.C.: 1971, *The American Political System: Notes and Readings* (revised ed.) Dorsey Press, Homewood, Illinois.

Busuttil, S., Agius, E., Serracino Inglott, P., Macelli, T. (eds.): 1990, *Our Responsibilities Towards Future Generations, A Programme of UNESCO and the International Environment Institute*, Foundation for International Studies, Valletta, Malta.

Callahan, D.: 1973, 'Science: Limits and Prohibitions', *Hastings Center Report* **3**(5), 5–7.

Davis, B.D.: 1991, *The Genetic Revolution: Scientific Prospects and Public Perceptions*, Johns Hopkins University Press, Baltimore, Maryland, London, U.K.

Engelhardt, H.T., Jr., 1996, *The Foundations of Bioethics*, 2nd ed., Oxford University Press, New York, NY, Oxford, UK.

Fletcher, J.: 1974, *The Ethics of Genetic Control: Ending Reproductive Roulette*, Anchor Press/Doubleday, Garden City, NY.

Gardner, W.: 1995, 'Can Human Genetic Enhancement Be Prohibited?', *Journal of Medicine and Philosophy* **20** (1), 65–84.

Golding, M.A.: 1978, 'Future Generations: Obligations to', *Encyclopedia of Bioethics*, The Free Press (Macmillan Publishing Co.), New York, NY, **2**, 507–512.

Heyd, D.: 1992, *Genethics: Moral Issues in the Creation of People*, University of California Press, Berkeley, California.

Hubbard, R., Wald, E.: 1993, *The Explosion of the Gene Myth: How Genetic Information is Produced and Manipulated by Scientists, Physicians, Employers, Insurance Companies, Educators, and Law Enforcers*, Beacon Press, Boston, Massachusetts.

Jonas, H.: 1984, *The Imperative of Responsibility*, University of Chicago Press, Chicago, Illinois.

Kavka, G.S.: 1982, 'The Paradox of Future Persons', *Philosophy & Public Affairs*, **11**(2), 93–112.

Knoppers, B.M.: 1991, *Human Dignity and Genetic Heritage [Dignité humaine et patrimoine génétique] (Protection of Life Series)*, Law Reform Commission of Canada, Ottawa, Canada.

Lappé, M.: 1978, 'Genetics and Our Obligations to the Future', in Bandman, E.L., Bandman, B. (eds.) *Bioethics and Human Rights: A Reader for Health Professionals*, Little Brown and Company, Boston, Massachusetts, pp. 84–93.

Lee, J.T., Klietmann, W., Ferrano, M.J.: 1995, 'Diagnosis and Detection of Drug-Resistant Strains of M. tuberculosis', *AIDS Clinical Care* (Massachusetts Medical Society) **7**(4), 27–29, 36.

Ludmerer, K.M.: 1978, 'Eugenics' [History], *Encyclopedia of Bioethics*, The Free Press (Macmillan Publishing, Co.), New York, NY, **1**, 457–461.

Margolis, J.: 1975, *Negativities: The Limits of Life*, Charles E. Merrill Publishing Co., Columbus, Ohio.

McNeill, P.M.: 1993, *The Ethics and Politics of Human Experimentation*, Cambridge University Press, New York, NY.

Muller, H.J.: 1935, *Out of the Night: A Biologist's View of the Future*, Vanguard Press, New York.

Muller, H.J.: 1958, *Man's Future Birthright* (an address at the University of New Hampshire, Nov. 21, 1957), University of New Hampshire, Durham, NH, pp. 1–24.

Nagel, E.: 1978, 'Comments on the Presentations of Drs. Ehrman and Lappé', in Bandman, E.L., Bandman, B. (eds.) *Bioethics and Human Rights: A Reader for Health Professionals*, Little Brown and Company, Boston, Massachusetts, pp. 94–97.

Parfit, D.: 1982, 'Future Generations: Further Problems', *Philosophy & Public Affairs* **11**(2), 113–172.

Parfit, D.: 1984, *Reasons and Persons*, Oxford University Press (Clarendon Press), Oxford, U.K., New York, NY.

Partridge, E.(ed.): 1981, *Responsibilities to Future Generations: Environmental Ethics*, Prometheus Books, Buffalo, New York.

Passmore, J.A.: 1974, *Man's Responsibility for Nature: Ecological Problems and Western Traditions*, Charles Scribner's Sons, New York, NY.

Pitkin, H.F.: 1967, *The Concept of Representation*, University of California Press, Berkeley/Los Angeles, CA.

Regan, T. (ed.): 1984, *Earthbound: New Introductory Essays in Environmental Ethics*, Temple University Press, Philadelphia, Pennsylvania.

Rolston, H.: 1988, *Environmental Ethics: Duties and Values in the Natural World*, Temple University Press, Philadelphia, Pennsylvania.

Salomon, M.: 1983, *Future Life* (trans. G. Daniels), Macmillan Publishing Co., New York, NY.

Schumpeter, J.A.: 1942, *Capitalism, Socialism, and Democracy* (2nd ed.) Harper & Brothers, New York, NY/London, U.K.

Sikora, R.I., Barry, B. (eds.): 1978, *Obligations to Future Generations*, Temple University Press, Philadelphia, Pennsylvania.

Suzuki, D.T., Knudtson, P.: 1989, *Genethics: The Clash Between the New Genetics and Human Values*, Harvard University Press, Cambridge, Massachusetts.

Watson, J.D., Juengst, E.T.: 1992, 'Foreword: Doing Science in the Real World: The Role of Ethics, Law, and the Social Sciences in the Human Genome Project', in G.J. Annas and S. Elias (eds.) *Gene Mapping: Using Law and Ethics as Guides*, Oxford University Press, New York, NY / Oxford, U.K.

Zimmerman, B.K.: 1984, *Biofuture: Confronting the Genetic Era*, Plenum Press, New York, NY, London, U.K.

UGO MIFSUD BONNICI

HOMO PROPHETICUS

It began with the formal definition of *homo sapiens*, and went through the description of *homo ludens, homo faber, homo ridens* or *cachinnans*, to *homo sapiens sapiens*. In fact a very distinctive trait in man is that of surmising a future: *homo propheticus*. Man is not only a historian with a personal and a collective memory, man is also a seer. In the case of the past, mankind's interest and wisdom is in the discovery of the truth. Very little can be done about the past except discovering it properly, and the ethical imperatives concern really the respect due to the truth. Regarding the future, there is a wealth of moral obligations, with perhaps a dearth of the possibilities of precise divination. Regarding the future we have responsibilities because we believe, more rightly than wrongly, that what the future will be, depends on us, individually and as a species.

It used to be said that man's presence had changed the physical structure and indeed the material reality of the world he inhabits rather marginally. From the moon, our planet's satellite, of all man's works, only one graffito, the Great Wall of China, is visible. All the great pyramids, imperial mounds and colosseums, wonderful and admirable within the short historical span dwindle into bare significance when considered on the cosmic scale.

No longer. We can pollute irreversibly our whole environment, remove mountains, conjoin oceans, tear open ozone layers. We have disturbed the balance of terror, whereby our revered species was kept within the numerical mean by disease, famine, earthquake and inundation, curbing over-reproduction.

Fifty years ago the circumstance of a global war offered the occasion for a demonstrative experiment with man's new destructive powers once nuclear energy came in hand. The effect has been to alert humanity to the threat from the new mastery of nature. Some misgiving opened the way to two opposite reactions: on the one hand the counter promethean syndrome took over, with a feeling that we were blasphemously daring beyond the limits of our wisdom, on the other hand the euphoria about the possibilities of subjecting further the kingdom of this earth over which we had been given dominion.

Perhaps not as garishly spectacular was the discovery of the helix of

E. Agius and S. Busuttil (eds.), Germ-Line Intervention and our Responsibilities to Future Generations, 165–168.

our DNA. Man seemed near to the secret of the transmission of genetic characteristics through a more intimate knowledge of the code of every living being. The consequences of this breakthrough could perhaps be even more far reaching. Teilhard de Chardin, scandalously for men of lesser faith, exulted at the new prospects for a step forward in the road of human evolution: man not as an ape of God, but man as a further discoverer not only of nature's laws but also of their keys.

The scientists who achieved the breakthroughs, somehow, had intimations of the new moral and commensurably heightened responsibilities. Unfortunately, this era of scientific discovery had been preceded by a period of time in which the systematic study as well as the stability of an ethical consensus had been under attack. Indeed Nietzsche had in fact proclaimed the death of morality and scorned its intellectual prestige.

Oppenheimer's scruples were deemed to be squeamish, and in the twilight zone of a well rooted but not well developed scientific ethic we have been feeling and groping our way through.

Is nuclear testing morally wrong? Is genetic engineering right? How far are we justified in advancing and what methods can or should be used. On the one hand, some would say that the discovery justifies the means; on the other, that man cannot and has never achieved progress through the sacrifice of human dignity or through the destruction of a single human personality, in the process of probing his way forward.

The primary obstacle to the formation of a consensus on the ethics involved in the process of genetic engineering is the desertification of our morality culture especially the part concerning the pre-birth period of human life, in addition to all connections with human sexuality and reproduction. One admires the brave attempt by Hans Küng and others to formulate what one would call "a global ethic," but the preliminary drafts afford in themselves ample evidence of the problem.

Fundamentally, the concept of the ethical imperatives as guidelines not impositions, as obligations whereby one binds himself to the stake of his own belief, not as chains clamped down by outsiders (indeed, of a religion as one's own freely embraced faith not as group tyranny) has not been firmly affirmed. Moral philosophy not as the philosophy of constraint – with good deference towards Professor Wildes' use of the term – but as the philosophy of restraint. Before ethics can form the basis of law and sanctioning by the community, there has to be a culture of inner persuasion.

Perhaps we should look towards the future and future generations not in a context of a break with a past and its cultural humus but as a challenge to our conscious – partly man-driven – evolution. The moral guide has to be worked at, perhaps with the same intellectual energy and academic co-ordination that the scientific world has shown in its dedication to research. One has to discount three background counter drives, the result of the last two centuries of history. First, one must not forget that for a considerable period of time there was a wide body of opinion which viewed science as a substitute to religion and as a "new" morality. This "scientism" is still very much alive, even though the wars and the excesses of the Nazi and other "scientific" regimes should have sobered us about this reliance on the scientific method as sole guide to practice.

In the second place, there exists the prejudice against all systems with built-in consistencies. This prejudice, originally part of Anglo-Saxon antipathy towards empirical wholly thought-out philosophies, has invaded with meandering and piecemeal pragmatic examination as a habit of mind, a good part of the academic milieu, world wide.

The third is the slightly misleading comprehension of the concept of pluralism. Pluralism does not, or should not, in my opinion, mean that any held position is as good or as bad as another. Pluralism implies the acceptance of a right to adhere to one's proper belief but does not bestow any charisma on the seperate individual philosophies of life. Pluralism is an abundance which can be a bounty in confluence, but would not be beneficient if it were taken to be a philosophical "qualunquism."

This collection of essays representing different cultural traditions can bear witness to:

1. the need for a morality to accompany science *pari-passu* in bio-genetics;
2. the fact that logic and consistency in thought are still required human qualities in arriving at moral conclusions; and the fact
3. that pluralism and diversity do not exclude confluence and synthesis.

Politicians and jurists may look into the matter of a formulation of laws to enforce what a community agrees should be enforced: definitely not your task, and it is important to distinguish between law and ethics (secular and religious) and morality (always somehow binding in conscience).

My wish is that the education of conscience through a profound study of the moral responsibilities contingent on our new powers and circumstances as the makers of the future of mankind, should be intensified, but should not hamper scientific research or dampen our urge to understand

more and to assist in the elimination or reduction of handicaps and genetic flaws, present in our fellow human beings, and could also burden
future generations. Given the basic dignity of human beings, no task
could be more noble.

NOTES ON CONTRIBUTORS

Emmanuel Agius, Ph.D., is Senior Lecturer in Ethics and Moral Theology, University of Malta; also, Coordinator of the Future Generations Programme, Foundation for International Studies, University of Malta.

Ugo Mifsud Bonnici is President of the Republic of Malta.

Salvino Busuttil is former Director-General of the Foundation for International Studies, University of Malta.

Alfred Cuschieri, Ph.D., is Professor and Head, Department of Anatomy and Director of the Institute of Gerontology, University of Malta.

Attajinda Deepandung, Ph.D., is Assistant Professor, Faculty of Social Sciences and Humanities, Mahidol University, Thailand.

H. Tristram Engelhardt, Jr., M.D., Ph.D., is Professor, Departments of Medicine, Community Medicine, and Obstetrics and Gynaecology, Center for Medical Ethics and Health Policy, Baylor College of Medicine; also, Professor of Philosophy, Rice University and Adjunct Research Fellow, Institute of Religion, Houston, Texas, U.S.A.

Alex E. Felice, Ph.D., is Professor and Head, Department of Pathology, University of Malta.

Louis Galea, LL.D., is former Minister for the Development of Social Policy, Malta.

David Heyd, Ph.D. is Professor of Ethics and Political Philosophy, Hebrew University, Jerusalem, Israel.

Kido Inoue is Zen priest, Master of the Geishu Tadanoumi, Shorinkutsu Seminary for Buddhist Priesthood, Japan.

Eric T. Juengst, Ph.D., is associate Professor of Biomedical Ethics, Case Western Reserve University School of Medicine, Cleveland, Ohio, U.S.A.

Tae-Chang Kim is President of the Institute for Integrated Study of Future Generations, Kyoto; Visiting Research Fellow at the Center for International Affairs, Osaka International University, Japan; also, Research Associate at the Center for Transformative Learning, Ontario Institute for Studies in Education, University of Toronto, Canada.

Wilai T. Noonpakdee is lecturer, Department of Biochemistry, Mahidol University, Thailand.

Qui Renzong, Ph.D., is Senior Research Fellow at the Institute of Philosophy, Chinese Academy of Social Sciences, and Professor of Philosophy of Science and Bioethics at the Graduate School of CASS.

Stuart F. Spicker, Ph.D., is Professor of Philosophy and Healthcare Ethics, Massachusetts College of Pharmacy and Allied Health Sciences, Boston, Massachusetts; Professor Emeritus, School of Medicine, University of Connecticut Health Center, Farmington, U.S.A.

Kevin Wm. Wildes, S.J., Ph.D., is Professor, Department of Philosophy, Georgetown University, Washington, D.C., U.S.A.

Katsuhiko Yazaki is Chairman of the Future Generations Alliance Foundation, Secretary-General of the Kyoto Forum, Japan, and Chairman of Felissimo Corporation.

INDEX

Philosophy and Medicine

23. E.E. Shelp (ed.): *Sexuality and Medicine*. Vol. II: Ethical Viewpoints in Transition. 1987   ISBN 1-55608-013-1; Pb 1-55608-016-6

24. R.C. McMillan, H. Tristram Engelhardt, Jr., and S.F. Spicker (eds.): *Euthanasia and the Newborn*. Conflicts Regarding Saving Lives. 1987   ISBN 90-277-2299-4; Pb 1-55608-039-5

25. S.F. Spicker, S.R. Ingman and I.R. Lawson (eds.): *Ethical Dimensions of Geriatric Care*. Value Conflicts for the 21th Century. 1987   ISBN 1-55608-027-1

26. L. Nordenfelt: *On the Nature of Health*. An Action-Theoretic Approach. 2nd, rev. ed. 1995   ISBN 0-7923-3369-1; Pb 0-7923-3470-1

27. S.F. Spicker, W.B. Bondeson and H. Tristram Engelhardt, Jr. (eds.): *The Contraceptive Ethos*. Reproductive Rights and Responsibilities. 1987   ISBN 1-55608-035-2

28. S.F. Spicker, I. Alon, A. de Vries and H. Tristram Engelhardt, Jr. (eds.): *The Use of Human Beings in Research*. With Special Reference to Clinical Trials. 1988   ISBN 1-55608-043-3

29. N.M.P. King, L.R. Churchill and A.W. Cross (eds.): *The Physician as Captain of the Ship*. A Critical Reappraisal. 1988   ISBN 1-55608-044-1

30. H.-M. Sass and R.U. Massey (eds.): *Health Care Systems*. Moral Conflicts in European and American Public Policy. 1988   ISBN 1-55608-045-X

31. R.M. Zaner (ed.): *Death: Beyond Whole-Brain Criteria*. 1988   ISBN 1-55608-053-0

32. B.A. Brody (ed.): *Moral Theory and Moral Judgments in Medical Ethics*. 1988   ISBN 1-55608-060-3

33. L.M. Kopelman and J.C. Moskop (eds.): *Children and Health Care*. Moral and Social Issues. 1989   ISBN 1-55608-078-6

34. E.D. Pellegrino, J.P. Langan and J. Collins Harvey (eds.): *Catholic Perspectives on Medical Morals*. Foundational Issues. 1989   ISBN 1-55608-083-2

35. B.A. Brody (ed.): *Suicide and Euthanasia*. Historical and Contemporary Themes. 1989   ISBN 0-7923-0106-4

36. H.A.M.J. ten Have, G.K. Kimsma and S.F. Spicker (eds.): *The Growth of Medical Knowledge*. 1990   ISBN 0-7923-0736-4

37. I. Löwy (ed.): *The Polish School of Philosophy of Medicine*. From Tytus Chałubiński (1820–1889) to Ludwik Fleck (1896–1961). 1990   ISBN 0-7923-0958-8

38. T.J. Bole III and W.B. Bondeson: *Rights to Health Care*. 1991   ISBN 0-7923-1137-X

39. M.A.G. Cutter and E.E. Shelp (eds.): *Competency*. A Study of Informal Competency Determinations in Primary Care. 1991   ISBN 0-7923-1304-6

40. J.L. Peset and D. Gracia (eds.): *The Ethics of Diagnosis*. 1992   ISBN 0-7923-1544-8

KLUWER ACADEMIC PUBLISHERS – DORDRECHT / BOSTON / LONDON